COMPENSATION GUIDELINES FOR ARCHITECTURAL AND ENGINEERING SERVICES

A management guide to cost-based compensation

Second Edition, Revised

A component of

THE AMERICAN INSTITUTE OF ARCHITECTS FINANCIAL MANAGEMENT SYSTEM

Endorsed by the American Consulting Engineers Council
1978
Published by The American Institute of Architects

First edition © 1975 by The American Institute of Architects was published under the title *Compensation Management Guidelines for Architectural Services*. This second edition, revised, © 1978 is published under the title *Compensation Guidelines for Architectural and Engineering Services.*

Library of Congress Catalog Card Number: 77-90943
ISBN 0-913962-03-1

Printed in the United States of America
AIA Catalog Number: 2M188 Rev. 1/78

Designer: Richard Darrell Sisk, The Associates, Inc.

Contents

Foreword

A major breakthrough was made in 1975 with the publication of *Compensation Management Guidelines for Architectural Services*. By setting down a rational process for relating design professionals' compensation to the cost of their services, some constraints to good client relationships were removed. The result: a process eminently fair and equitable to both.

The need for greater simplicity, comprehensiveness and uniformity of detail has led to this rewritten second edition, titled *Compensation Guidelines for Architectural and Engineering Services* to reflect its broader concerns. It is one component of the just emerging *AIA Financial Management System*. Taken together, the family of components should contribute to greater financial precision in the business practices of design professionals. By 1979 all the parts are expected to be in place, but each will be made available as it is developed.

Briefly, the AIA Financial Management System as currently envisioned is made up of the following components:

—FM Principles
—FM Manual Applications
—FM Automated Applications
—FM Compensation Guidelines
—FM Time Data

The first component, "Principles," in the form of a *Financial Management Handbook*, will set down basic theories and applications appropriate to firms of design professionals, as well as acting as an umbrella for the system. The manuscript, just getting under way, also provides for inclusion of pertinent information drawn from several earlier AIA publications which the new handbook will replace.

The keystone of the second component, "Manual Applications," is a new book on *Standardized Accounting for Architects*, now in press. Together with its interrelated accounting forms, it replaces an out-of-print publication of the 1950s. The all-new book and forms, which spell out uniform manual accounting procedures specially designed for small and medium-size offices, is coordinated with the automated mode for easy transfer at the proper time.

In use since 1971, the AIA *Computer-based Financial Management Service* (CFMS) has recently been modified to provide the detailed in-house data recommended for use with *Compensation Guidelines*. Together with its users' manuals, input forms and output printouts, CFMS is the third component of the AIA Financial Management System, "Automated Applications." Available by subscription to architectural, engineering and other professional firms, CFMS provides a comprehensive project control system and a fully integrated accounting system. This service is available by mail or by teleprocessing to local computer terminals.

This book, with its related forms and documents, makes up the fourth component, "Compensation Guidelines." Also available are two supplementary publications, one especially written for clients, the other for architects and engineers. A thorough understanding of the concepts and procedures set forth in *Compensation Guidelines for Architectural and Engineering Services* will also make the other components much easier to understand and use.

The final component, "Time Data," describes the companion to *Compensation Guidelines:* the *AIA Time Data Bank*. Currently under development as a national AIA program, it is aimed at sharing among members the amounts of time required to provide items of professional services for a diversity of recent projects. Its purpose is to provide an accessible data base for checking personnel time requirements along with estimating compensation for new projects.

By embracing all components of the AIA Financial Management System, users will be able to sharpen their financial and management skills to add to their design competence.

JOHN M. MCGINTY, FAIA
President, The American Institute of Architects

Washington, DC
December 1, 1977

Preface

COMPENSATION MANAGEMENT

Society expects good design and a businesslike approach from its architects and engineers. A rational, consistent and accountable system for determining and managing compensation is therefore as important to the architect/engineer (A/E) as it is to the client.

For years architects and engineers have struggled with ill-adapted and often unworkable procedures for determining their compensation. Perhaps the most troublesome of these procedures has been the one using a percentage of the construction cost, because it rewards an effort to *reduce* construction cost by *reducing* the A/E's compensation.

Furthermore, the services offered by the A/E have expanded far beyond those traditionally required. Also, many clients have developed their own requirements for determining compensation.

To accommodate these new circumstances, an approach to compensation has been developed to quickly and accurately identify and itemize services required to carry out a project, assign responsibility, and estimate the costs to provide such services.

The approach also sets up a consistent vocabulary of terms and procedures. It can serve as a basis of discussion with the client; and, used to its fullest, it will help the A/E manage a project and monitor its costs.

It is not a *method* of compensation, but a guide to determining the *amount* of compensation appropriate to the services provided, based on an estimate of the A/E's costs. *Any of several methods of compensation may be used.*

The approach has been in use since early in 1975, and these *Guidelines* are designed to show the A/E and client how to work with it.

A first edition of the *Guidelines* was issued by The American Institute of Architects in February 1975 with the help of a blue-ribbon task force, incisive comments from a representative group of clients and allied disciplines, earlier work developed by several components of AIA, and the keen efforts of staff at AIA Headquarters. This second edition is the result of a review process that involved feedback from owners and practitioners who used the guidelines in practice.

AIA FINANCIAL MANAGEMENT SYSTEM COMPONENTS

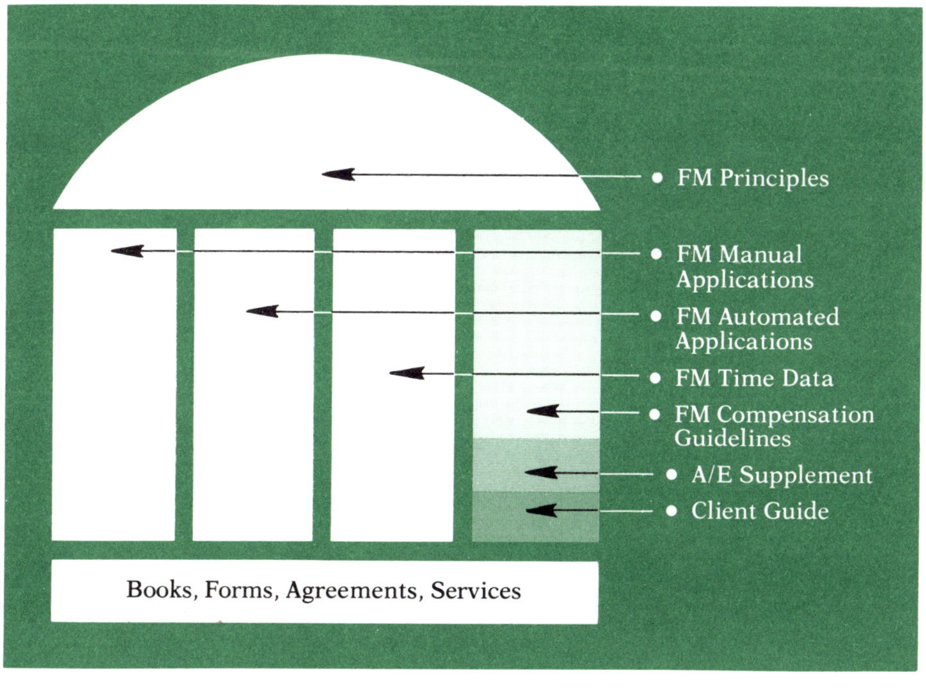

THE AIA FINANCIAL MANAGEMENT SYSTEM

The *Guidelines* should be seen in the context of a broad Financial Management System (FMS) under development by the AIA (see accompanying illustration and the Foreword).

ARCHITECT—ENGINEER LIAISON

The AIA worked closely with the American Consulting Engineers Council (ACEC) in gearing this second edition of the *Guidelines* to the common concerns of the members of both societies. For example, the listing of services has been improved as a result of this collaboration. Similarly, Chapter 5 refers to the agreement forms issued by both societies.

The liaison with ACEC which produced this rewritten edition is planned to continue in the years ahead on a broadened base. For example, under consideration is a detailed list of services offered by each engineering discipline when the architect is the prime professional on a project; so is a parallel list when the engineer is the prime professional.

RESPONSE AND IMPROVEMENTS

Readers of these guidelines and users of the compensation management component of the AIA Financial Management System are encouraged to send in any comments and suggestions they may have. These should be sent to Robert Allan Class, AIA, at The American Institute of Architects, 1735 New York Ave., N.W., Washington, DC 20006.

The A/E's main asset is people. Most of the cost of rendering professional services goes for paying the people who do the work. The *Guidelines* recognize this principle.

ACKNOWLEDGMENTS

The following contributed to the development of the second edition of the *Guidelines:*

AIA Task Force on Compensation Management
Ward W. Deems, FAIA (Chairman), San Diego
Gerald R. Clark, AIA, Phoenix
James A. Greene, FAIA, Tampa
E. Keith Haag, FAIA, Cuyahoga Falls, OH
Robert L. Halford, AIA, Dallas
George W. Sprinkle, FAIA, Phoenix
Gary S. Van Neck, AIA, Detroit
William R. Wallace, AIA, Bethlehem, PA

AIA Board of Directors
Donald J. Stephens, FAIA, Albany, NY

AIA Office Management Committee
Henry W. Schirmer, AIA (Chairman), Topeka

AIA Task Force on Financial Management
Peter Piven, AIA (Chairman), Philadelphia

AIA Documents Board
Edward D. McCrary, AIA (Chairman), San Francisco

AIA Staff
Robert Allan Class, AIA (Adviser, and Editor of 1st edition), Washington, DC

ACEC Guidelines Task Force Committee
Michael A. Hertzberg, PE (Chairman), Warren, VT
George C. Frost, PE, Glen Falls, NY
Preston E. Newman, PE, Atlanta
E. N. Nicolaides, PE, Miami
Thomas Carcaterra, PE, Silver Spring, MD
Bruce C. Roberts, ACEC Staff, Washington, DC

Editor
Stephen A. Kliment, AIA, New York

The above positions represent appointments made in 1976.

1. Introduction and Summary

The continuing expansion of services provided by the architect/engineer (A/E) includes new and complex aspects of traditional services, as well as a growing involvement with energy conservation, environmental protection and preservation, pollution control and land use policies.

To accommodate this broadened spectrum, as well as a greater need for financial precision, the architectural and engineering professions have collaborated to develop a logical and direct process for determining the scope of their clients' needs, along with a positive way for determining adequate and mutually satisfactory compensation for professional services.

ELEMENTS OF COMPENSATION MANAGEMENT

The approach to compensation management described in these *Guidelines* consists of:

—A standardized listing, identification and description of those professional services which may be provided by architects and engineers in designing the built environment.

—A method for:
- —determining, in concert with the client, the services needed for the client's project;
- —allocating responsibilities for these services among the design professionals and client;
- —determining proper compensation for professional services;
- —assisting the A/E in monitoring office costs during service;
- —estimating costs for services on similar future projects;
- —assisting in revising A/E services to account for changes in project scope.

The approach allows client and A/E to gain a full knowledge of the elements that make up the A/E's estimated costs and planned profit *prior* to signing an agreement for professional services. At the same time, the client is able to obtain a precise understanding of the broad range of services the A/E has to offer.

EXHIBIT 1-1. STEPS IN CARRYING OUT THE GUIDELINES APPROACH TO COMPENSATION MANAGEMENT

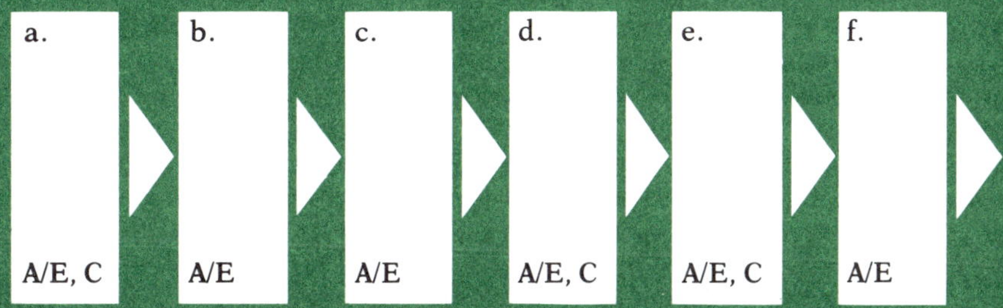

a. Determine scope of designated services
b. Establish hours and costs
c. Summarize compensation by phase and project
d. Determine amount, method of compensation
e. Prepare, sign agreement
f. Establish in-house project control

A/E = by Architect/Engineer
C = by Client

PROCEDURES

The following steps are recommended when using the approach (see Exhibit 1-1):

Step 1. Client and A/E jointly determine and record each item of the scope of designated services required for the project and the specific responsibility of the client, A/E or consultant for each item of service (see Chapter 3).

A wide range of services available through architectural/engineering practitioners has been classified into nine categories of service (see Appendix A). The five traditional sequential phases (Schematic Design, Design Development, Construction Documents, Bidding or Negotiations, and Construction Contract Administration) have been supplemented with three additional phases (Predesign, Site Analysis, and Postconstruction). A nonsequential phase entitled "Supplemental Services" is provided for those services usually not restricted to a particular sequential phase.

From a series of service lists, the client and A/E jointly determine the scope of designated services required for the project and the specific responsibility of the client, A/E or consultant for each item of service.

Step 2. The A/E estimates the number of direct hours for providing each item of service or group of services, the cost of providing each such item or group, and the amounts of proposed compensation and anticipated reimbursable expense (see Chapter 4).

A clear understanding of the work required under each item of service is essential if the A/E is to estimate the working time for each such service item. Time estimates are the heart of the cost-based compensation approach.

Step 3. The A/E summarizes for each phase and for the project the amounts of proposed compensation and anticipated reimbursable expense (see Chapter 4).
Based on estimated time requirements, a tabulation is made of required compensation. For each project, the A/E follows regular accounting procedures to determine all elements of expense and profit which must be accounted for within the compensation.

Step 4. Using the written results of steps 1 and 3 as aids, client and A/E jointly determine the amount and method of compensation, and the description of each item of service required (see Chapter 4 and Appendices A, B and D).
Flexibility in establishing methods of compensation for various types and phases of projects is a must. Any of the methods described in Appendix B or variants and combinations thereof may be employed once the A/E has estimated, based on the firm's projected costs, the level of compensation required to carry out the agreed-upon services.

Step 5. The client and A/E arrange for preparation and then sign an appropriate agreement for professional services.
AIA and ACEC offer documents and contractual aids which provide legal clarity and simplify the process of reaching an agreement. They set out standard terms and conditions which normally do not vary from project to project and have proven equitable to all parties over years of use. A brief discussion of several agreement forms available from these professional societies appears in Chapter five. In addition, the new AIA Documents B161 (Standard Form of Agreement Between Owner and Architect for Designated Services) and B162 (Scope of Designated Services) are included in Appendix D.

Step 6. The A/E can then establish inhouse project control (Chapter 6).
Now that the A/E firm has the scope and estimated time and cost of services, it has the groundwork for its own timely internal project control system. Monitoring job cost against original estimates on the project will let the firm know in good time when costs encroach on profit.

By following these steps, the A/E is able to relate compensation directly to the time estimated for rendering the specific services agreed upon with the client.

THE APPROACH NEED NOT BE COMPLEX

The *Guidelines* are designed to be used at several levels of complexity. On a large, complex project, the A/E will normally review the scope of designated services and then do a detailed compensation analysis by service and phase to arrive at an accurate estimate of compensation. On a simpler project, after first determining the scope of designated services, the A/E may do the analysis by phase only. Chapters three and four include examples of both the "long-form" and "short-form" applications of the *Guidelines*.

As the A/E's services become broader and more complex, the approach described in the *Guidelines* will not only serve existing needs but readily adapt themselves to the many still undetermined services of the future.

2. Compensation Principles and Definitions

The following accounting and related terms and definitions were developed with the help of professional accountants and computer consultants, as well as A/E practitioners with many years of experience in managing projects and offices. Note that federal agencies usually create and follow their own definitions and must be consulted when preparing project cost estimates for such clients.

While the definitions and commentary in this chapter are not intended to reflect all accounting principles and procedures, the discussion has sufficient depth to set the stage for proper use of the forms presented later in the *Guidelines*.

COMPENSATION PRINCIPLES

Payments that flow from the client to the A/E for professional services rendered consist of *compensation* and *reimbursable expense*. Compensation may be considered as divided between *expense* and *profit*. Expense (a term synonymous with "cost") is made up of direct (project) expense and indirect (overhead) expense, the latter to be allocated on a pro rata basis among all current projects. Direct expense may include personnel expense, outside services (e.g., consultants') expense, and other expense (e.g., reproduction of in-house documents). Indirect expense also includes certain personnel expenses (e.g., promotional or office management time) and other expenses.

Reimbursable expense for out-of-pocket expenses incurred by the firm on behalf of the project is paid for separately from compensation. These items are as cited in the agreement for professional services and may include reproductions, transportation, in-house documents, living expenses, postage and handling. Whether an item is assigned as a direct project expense or a reimbursable expense must be determined by the A/E and client.

Some A/E firms place a markup on reimbursable (and certain direct) expenses to cover the cost of their administration and financing.

Profit is defined as the excess of income finally earned over and above the A/E's incurred expenses. These expenses include principals' salaries. It is that element of compensation which provides reward for professional and financial risk, and return on investment; it is a major incentive for being in business. When *planning* profit, an allowance should also be considered to cover certain unbillable items, uncompensated increases in indirect expenses after the indirect expense factor has been determined, and fee for assuming responsibility for consultants' services (which amounts to a markup on their charges).

The amount of profit planned for a project should relate not only to the circumstances of that project, but to the overall financial objectives of the firm. Planning income, expenses and profit makes up the *profit plan*.

Actual profit may or may not exactly equal planned profit. Poor project management, estimating errors, unexpected cost increases, uncontrollable project delays and other factors often cause actual profit to fall short of the amount planned.

As noted in the chapter "Financial Management" of *Current Techniques in Architectural Practice* (see Reference 5, Appendix C), "The essential concept in the profit plan is the interrelationship between total income and total expense and, more particularly, between net income from projects (also known as gross margin) and indirect expense. It is commonly accepted that profit is earned when gross income exceeds gross expense. In the architectural profession, income is largely a consequence of project compensation, and consequently, profit is earned when the net income from projects—the amount left after direct expenses—exceeds the firm's indirect expenses."

Proper accounting procedures and record keeping are essential for realizing the full benefits of this approach. Any firm which has, at a minimum, an elementary bookkeeping system with project expense accounting can easily use the *Guidelines* without having to add procedures. If time cards have been retained—even though records have not been compiled from them—the firm will have a data base it can refer to. By using data judiciously gathered from a variety of sources as the basis for estimates, even the very small or simply organized firm may successfully employ these principles.

In computing profit or loss, the charts-of-account sequence has a direct bearing on the form of the Income and Expense Statement. This Statement, in turn, yields data used in computing Benefits and Indirect Expense Factors when determining compensation on new projects. The following sequence is used throughout the AIA Financial Management System:

Income
Less Reimbursable Expense
Subtotal
Less Direct Expense
Subtotal ("Contribution")
Less Indirect Expense
Profit (Loss)

Placing Reimbursable Expense immediately after Income allows the reimbursables to be offset. Subtracting Direct Expense from the Income subtotal produces the "Contribution" to Indirect Expense and Profit.

Approaches may differ among A/E firms when computing the elements of proposed compensation, due to their respective preferences, accounting procedures or clients' requirements. Agree-

EXHIBIT 2-1. PAYMENTS TO THE A/E FOR PROFESSIONAL SERVICES

Compensation			
Expense			Profit
Type ▶ ▼	Direct Expense	Indirect Expense	
Personnel Expense (Principals and Employees) ▶	Direct Personnel Expense (Direct Salary Expense plus Benefits)	Indirect Personnel Expense (Indirect Salary Expense plus Benefits)	
Outside Services Expense ▶	Direct Outside Services Expense	Indirect Outside Services Expense	
Other Expenses ▶	Other Direct Expense (Nonreimbursable)	Other Indirect Expense	and Contingency

Reimbursable Expense

ment on basic definitions and procedures will tend to reduce the impact of such variations, however. Simplified definitions that identify the elements of compensation and reimbursable expense have been developed for use with the *Guidelines*. The user must understand these terms and the links between them in order to interpret the directions and comments contained in later chapters.

DEFINITIONS, WITH COMMENTS

Compensation is payment received or receivable for services rendered. Compensation is used first to pay *expense* incurred; what is left is *profit*. (See Exhibit 2-1.)

Expense is payment made or payable by the firm for goods, services and other obligations allocable to current operations and is synonymous with "cost." Expense, the major portion of compensation, is either *direct* (project-related), *reimbursable*, or *indirect* (overhead). It is made up of various kinds of expenses, including personnel, outside services, and other expenses. Reimbursable expenses are not considered a part of compensation.

Personnel Expense is the sum of salary and benefits.

Salary (a part of Personnel Expense) is gross pay (not including bonus and profit sharing) for services rendered to the firm by employees and principals. *Only gross pay* should be considered as salary when applying the *Guidelines* and determining hourly rates (see also *Direct Personnel Expense* and *Direct Salary Expense*). Working Time Salary is Total or Gross Salary minus the cost of Paid Time Off; this is used in determining Indirect Expense Factor as shown in Exhibit 2-5.)

Direct Expense is all expense allocable to a specific project (see Exhibit 2-2).

Direct Salary Expense—a part of *Personnel Expense*—is principals' and employees' gross pay (excluding direct benefits, bonuses and profit sharing) attributable to projects. *Indirect salary expense*, also a part of personnel expense, is

EXHIBIT 2-2. EXAMPLES OF DIRECT (project-related) EXPENSE

These items cannot always be estimated with accuracy or made to fit within statutory requirements. Also, some of the items are best handled as reimbursable expense items, as provided for in the agreement for professional services. In cases where this is not possible, they should be carried as direct expense.

DIRECT PERSONNEL EXPENSE
Salaries and benefits: employees and principals working on project

DIRECT OUTSIDE SERVICES EXPENSE
Fees for project services provided by consultants

OTHER DIRECT EXPENSE

Reproduction of drawings and specifications

Printing and duplication

Photographs and microfilms

Transportation

Living expenses other than at home

Long distance telephone, telegraph, telex, etc.

Postage for mailing drawings and specifications

Premiums for professional liability insurance written for project

Premiums for professional liability insurance required for project in excess of basic coverage normally carried by firm

Other project-related insurance premiums

Legal and accounting services

Data processing

Meals during meetings

Items paid for on client's behalf, such as: fees, permits, bid advertising, models, renderings. Implied responsibility could result from the A/E paying for items such as surveys, borings, soil investigations; the client should either contract directly for such services or execute an agreement that the A/E is neither providing nor guaranteeing such services.

Financing and "carrying" costs of professional services at client's request. In effect, this is akin to the A/E making a loan to the client.

gross pay *not* directly allocable to specific projects (see also *Salary, Indirect Personnel Expense*).

Benefits (a part of personnel expense) are defined as additions to salary paid by the firm on behalf of its personnel. Examples of "mandatory and customary" benefits are mandatory employer's payroll taxes (as employer's share of Social Security contributions, unemployment taxes, Workers' Compensation insurance), and customary employer's contributions to insurance and pension programs, as well as paid time off for such as vacations, holidays and sick leave. The term "Direct Benefits" refers to that portion of Total Benefits that is in the same proportion that Direct Salaries are to Total Salaries. No matter how the costs of these benefits are treated as an *accounting* item, it is usual from a *compensation* viewpoint to include the pro rata share of these items (i.e., Direct Benefits) in Direct Personnel Expense as illustrated in Exhibits 2-3 and 2-5.

Some firms customarily (and sometimes selectively) provide additional benefits or perquisites (such as unlimited use of company car, parking space, vacation expense, education, medical costs not covered by health insurance). These may be disallowed as a personnel expense by some private clients and local, state and federal agencies. In such cases, the items should be considered in planning profit when estimating compensation. Bonuses and profit sharing are not considered benefits within the meaning of the *Guidelines;* they are allocated to other accounts, depending upon each firm's accounting procedures.

Benefits Factor is the cost of mandatory and customary benefits (including paid time off) expressed as a percentage or multiple of total salary expense for working time or as an hourly rate. This

EXHIBIT 2-3. DETERMINING BENEFITS FACTOR—AN EXAMPLE

EMPLOYEE PAY AND BENEFITS SUMMARY

SALARY BASED ON 40-HR. WEEK:	ANNUAL	$16,640
	WEEKLY	320
	DAILY	64
	HOURLY	8
BENEFITS:		
MANDATORY (FICA, Fed. & State Unemployment Insurance, Workers' Comp.)		$ 1,046
CUSTOMARY (Health, Life and Disability Insurance Premiums, Pension Plan, AIA Dues)		714
PAID TIME OFF (30 days per year for Vacation, Holidays, Sick Leave, Emergency Leave = 240 hrs. @ $8/hr.)		1,920
TOTAL COST OF BENEFITS		$ 3,680
ACTUAL WORKING TIME = 2080 − 240 = 1,840 Hrs.		
TOTAL COMPENSATION = $16,640 + $1,046 + $714 =		$18,400
COMPENSATION PER HOUR WORKED = $\frac{\$18,400}{1,840}$ =		$ 10

$$\text{BENEFITS FACTOR} = \frac{\text{Total Cost of Benefits (Mandatory/Customary + Paid Time Off)}}{\text{Total Direct + Indirect Salary Expense (excluding Paid Time Off)}}$$

$$= \frac{\$3,680}{\$8 \times 1,840 \text{ hrs.}} = \frac{\$\ 3,680}{\$14,720} = 25\%$$

NOTE: The word "Total" in the Benefits Factor formula means all defined expense for the term of the period selected. This period should be the same for all computations. The period most frequently selected is the firm's fiscal year. This example is for a single employee. The same formula is used to calculate the average benefits factor for the firm.

factor may be obtained by analyzing the firm's records relating to mandatory, customary and additional benefits. The simplest approach is to determine the factor as an average for all personnel. The A/E should, however, recognize variations in benefits for various levels of personnel when breaking down estimated costs by personnel level or individuals. See Exhibit 2-3 for an example of a typical computation.

Direct Personnel Expense (sometimes known as "DPE") is defined as the direct salaries of all personnel (including professional, technical and clerical) engaged on a specific project, and the portion of the cost of their mandatory and customary benefits related thereto. In project cost accounting, include principals at an appropriate rate as compensation for their time and effort expended directly on the project. Additional payments to principals beyond their minimum acceptable project-related compensation are charged to other accounts.

In the *Guidelines,* the term "principals" is broadly defined as those who have an equity interest in the firm—whether a sole owner of a proprietorship, partners in a partnership or officers of a corporation. The A/E using the *Guidelines* needs to determine the specific personnel in the firm who are classified as principals.

Direct Outside Services Expense is made up of payment for services rendered to the firm by outside, independent service contractors or professional consultants.

Other Direct Expense covers payment for all of the firm's nonreimbursable direct expenses, except for personnel and outside services expense. It can include reproduction of drawings and specifications for in-house use, travel expenses, long distance communications, and items paid for on the client's behalf, if not reimbursed. (See Exhibit 2-2.)

Indirect Expense is defined as all expense not directly allocable to specific projects and is synonymous with "overhead." Examples of indirect expense are shown in Exhibit 2-4. (See the definition and discussion of *indirect expense factor,* below, for the method of allocating a pro rata portion of indirect expense to a specific project.)

Indirect Personnel Expense is the sum of indirect (non-project-related) salary expense and indirect (non-project-related) benefits expense.

Indirect Expense Factor is the cost of all indirect expenses, including indirect personnel expense, expressed as an hourly rate or as a percentage or multiple of either Direct Salary Expense or Direct Personnel Expense, depending upon how a firm applies benefits expense to projects.

EXHIBIT 2-4. EXAMPLES OF INDIRECT (non-project-related) EXPENSE

Some of the items below also appear in examples of direct expense; the test for allocation of expense between direct and indirect is whether or not such items are project-related. (Some items of indirect expense may be disallowed by certain clients and government agencies in calculating indirect expense factor; in such cases, these items should be allocated as part of planned profit.)

INDIRECT PERSONNEL EXPENSE

Salaries and benefits: employees and principals working on non-project-related tasks

INDIRECT OUTSIDE SERVICES EXPENSE

Fees to certain consultants, such as legal, accounting, public relations, personnel, management, etc.

OTHER INDIRECT EXPENSE

Capital expenditures

Depreciation: typically on furniture, equipment, building, automobiles
Amortization: typically on leasehold improvements

Prepaid expenses

Expenses having an effect on operations beyond current fiscal year and required by Internal Revenue Service to be spread over more than one year, such as long-term insurance premiums, certain taxes, equipment leases, licenses/fees, organization expenses

Current operating expenses

Rent (or equivalent), utilities, operation, maintenance, repair, cleaning: for building, equipment, automobiles
Duplication, photographs, microfilms, data processing
Postage and messenger
Travel, entertainment (promotional)
Office and drafting supplies, books and periodicals
Telephone, telegraph, telex, etc.
Taxes, such as real estate, personal property, state franchise (for corporations)
Professional dues, registration fees, seminar and convention expenses, licenses
Contributions, charitable and civic
Insurance, such as automobile, contents, building, principals' life, valuable papers, employer's liability, equipment, professional liability
Interest on such as auto, equipment, operating loans
Development, promotion, public relations
Bond premiums

Losses

Such as from thefts or casualty not covered by insurance, forgeries or embezzlement not covered by bond, uncollectable compensation, bad debts, deductibles on insurance payouts

To determine this factor, one must analyze the firm's records relating to all items of indirect expense incurred during a selected period. The data should reflect the most current expense items and requires careful accounting. The most common way to determine the factor is to use historical data from a past accounting period, such as the previous fiscal year. A more accurate method is to use data from the four most recent quarters or, even better, from the twelve most recent months. (See Exhibit 2-5.)

After a base accounting period is selected, all items of indirect expense for the period are totalled, as are all items of direct personnel expense (or direct salaries) for the same period. The relationship between the two becomes the indirect expense factor. This may be expressed as a multiple or percentage (ratio), or it may be converted to an hourly rate.

In determining Indirect Expense Factor for purposes of estimating compensation, the factor may be related either to Direct Salary Expense or to Direct Personnel Expense. The latter includes Direct Benefits, while the former does not. When relating to Direct Salary Expense, it is essential that Direct Benefits be accounted for. The best way to do this is to add Direct Benefits to Indirect Expense, as indicated in the example in Exhibit 2-5. The cost of Direct Benefits must be recovered, regardless of the method used.

Even though it is simplest to determine the factor as an average for *all* personnel, note possible variations in indirect expense factor that apply to various levels of personnel when breaking down estimated costs by personnel level or individual.

Profit is what remains after all expenses have been accounted for; it is a positive arithmetical difference between income and expense. *Loss* is a negative

EXHIBIT 2-5. DETERMINING INDIRECT EXPENSE FACTOR—AN EXAMPLE

Apple and Bartlett Associates (see Exhibit 4-2) is a hypothetical architectural/engineering firm with a total staff of 12, a projected gross annual payroll of $195,520, a projected annual salary expense for working time of $172,960, an expected cost of benefits (including paid time off) of 25% of working time salary expense, and a budget for indirect expense (excluding indirect salary and benefits) of $86,135. For planning purposes it is projected that 76% of working time salary expense will be for project-related activities and the remaining 24% for non-project-related tasks. The following is an outline of a portion of the firm's expense budget; *not* shown are DIRECT OUTSIDE SERVICES EXPENSE, OTHER DIRECT EXPENSE or PLANNED PROFIT.

ITEM	DIRECT EXPENSE	INDIRECT EXPENSE	TOTAL
Working Time Salary Expense	$131,450 (a)	$ 41,510 (b)	$172,960 (c)
Benefits, including paid time off (25%)	32,862 (d)	10,378 (e)	43,240
Non-Personnel Indirect Expense		86,135 (f)	86,135
Totals	$164,312 (g)	$138,023 (h)	$302,335

Based on the preceding:

Working Time Salary Expense	$172,960 (c) or (a + b)
Direct Salary Expense	$131,450 (a)
Direct Benefits	$ 32,862 (d)
Direct Personnel Expense	$164,312 (g) or (a + d)
Indirect Expense	$138,023 (h) or (b + e + f)

In determining compensation INDIRECT EXPENSES are recovered by use of an INDIRECT EXPENSE FACTOR; two ways are illustrated:

- As a multiple (or percentage) of DIRECT **SALARY** EXPENSE:

$$\text{INDIRECT EXPENSE FACTOR (based on Direct Salary Expense)} = \frac{\text{Total Indirect Expense (\textbf{including} Direct Benefits)}}{\text{Total Direct \textbf{Salary} Expense}}$$

$$= \frac{(h + d)}{(a)} = \frac{138{,}023 + 32{,}862}{131{,}450} = \frac{170{,}885}{131{,}450} = 1.30\ (130\%)$$

- As a multiple (or percentage) of DIRECT **PERSONNEL** EXPENSE:

$$\text{INDIRECT EXPENSE FACTOR (based on Direct Personnel Expense)} = \frac{\text{Total Indirect Expense (\textbf{excluding} Direct Benefits)}}{\text{Total Direct \textbf{Personnel} Expense}}$$

$$= \frac{(h)}{(g)} = \frac{138{,}023}{164{,}312} = 0.84\ (84\%)$$

See note, Exhibit 2-3, for definition of "Total"

arithmetical difference between expense and income. See previous discussion in this chapter under "Compensation Principles."

Contingency, though not an accounting item, is an allowance used in estimating compensation to cover such unknowns as possible project complexities, minor changes in project scope, minor project delays or other unforeseen conditions. Whether or not the A/E includes this factor is strictly a matter of judgment. A reasonable contingency factor should probably be applied to all hourly estimated items when using the *Guidelines*.

Reimbursable Expenses are out-of-pocket expenses incurred by the A/E on behalf of a project. They are paid back to the A/E separately from compensation. These expense items are generally not directly related to the actual providing of professional effort.

3. Identifying Services

The first of the several steps in determining cost-based compensation—once the client has been apprised of its main features and requirements—is for the client and A/E to determine jointly and to record each item in the scope of services required or not required for the project. Responsibility is then assigned for each item to the client or the A/E. When reference is made to terminology used in AIA forms, the term "architect" will be used in lieu of "A/E," and the term "owner" will be used in lieu of "client," as these are the terms used in AIA Owner-Architect agreement forms.

To help complete this step, note a typical "Schedule of Designated Services Worksheet" form (Exhibit 3-2). (The full set of forms—AIA Forms F810-819—is included in Appendix D at the end of the *Guidelines*.) Forms F811-819 are preprinted—one for each of the phases into which a project is commonly divided, thus assisting in establishing a professional standard. In Form F810, the wide space at the left of the form is blank and may be used for simpler projects whose scope of work is best identified by phase alone; or several phases may be combined; or it may be used for a limited or special combination of services. Before proceeding to fill out the Schedule of Designated Services forms, the A/E ("architect") and client ("owner") should review the scope of designated services schedule (Appendix A) and *descriptions* of the items of service (Appendix D, as part of AIA Document B162).

Note that the descriptions of items of service in B162 are *general* indications of what may go into each item. Review them carefully for each project, and modify them as appropriate to make them *specific* to the project by deletions and/or additions to the descriptions. Note that some descriptions contain multiple choices, some of which must be deleted.

The purpose of the *Schedule of Designated Services Worksheet* forms (F810-819) is to help determine the specific services required for a project and allocate to either A/E or client the responsibility for providing each applicable service. Determination and allocation are made jointly by A/E and client and recorded on the form.

Follow this procedure in completing the Schedule of Designated Services forms:

—Use the first two columns on the form to allocate responsibility for services to be provided by the A/E. The column "By Architect" is used to mark services to be provided in-house by the A/E. The column "By Architect, as Outside Services" is used to mark services to be provided by the A/E through outside consultants; different symbols may be used to designate different types of consultants. These are services for which the A/E will assume responsibility. The column "By Owner, Coordinated by Architect" is used to mark services which the client will provide and which will require some coordination by the A/E in the interests of the project. The column "By Owner" is used to mark services to be provided by the client exclusively and in which the A/E will not participate in any way.

—The column "By Architect, as Additional Service" serves as a notice that the item has been discussed, method of compensation determined, and authorization to provide services will be given in writing at a future date.

—In the column "Not To Be Provided," mark on the preprinted form the listed services not required for the project.

—Use the blank (vertical) columns to allocate responsibilities to specific parties responsible to the A/E or client in other combinations than those indicated in columns 1, 2, 5 and 6 (for example, lessees, construction manager, using agency, etc.).

—Place a check mark after each listed service in at least one of the columns identified with a heading. Shared responsibilities can be indicated by marking more than one column. Symbols can be used to indicate primary and secondary responsibilities. As each listed item of service is marked, be sure to understand (and modify as necessary) its description as it appears in AIA Document B162. This is vital in reaching a mutual understanding of the work to be included or excluded.

—Use the blank (horizontal) lines in the lower part of the preprinted forms to enter services required for the project but which are not listed. A description of

EXHIBIT 3-1. PROJECT CONDITIONS USED FOR COMPLETING EXHIBIT 3-2

The project is an elementary school. The budget is $1.85 million. The hypothetical A/E firm of Apple & Bartlett Associates has been commissioned to provide professional services in Phases 3 (Schematic Design); 4 (Design Development); 5 (Construction Documents); 6 (Bidding or Negotiations); and 7 (Construction Contract Administration).

In addition to Architectural work, all basic Civil and Structural Engineering will be done in-house, with Mechanical and Electrical to be done by consultants. Landscape design and installation will be provided by the City Parks Department and coordinated by the A/E. Interior Design by the A/E will be for cabinet work and placement of school equipment. Due to program requirements for "open plan," special Graphics design is required.

Bidding will be for traditional construction sequence, not "fast-track." Four prime construction contracts will be let.

EXHIBIT 3-2. SAMPLE COMPLETED SCHEDULE OF DESIGNATED SERVICES WORKSHEET FORM F815

SCHEDULE OF DESIGNATED SERVICES WORKSHEET — **PHASE 5: CONSTRUCTION DOCUMENTS SERVICES**

Project *Franklin Elementary*

Project # *7701* Date *1·2·77*

Owner *Alton School Board*

Architect *Apple & Bartlett Assoc.*

	1 By Architect	2 By Architect, as Outside Services	3	4	5 By Owner, Coordinated by Architect	6 By Owner	7	8	9 By Architect, as Additional Service*	10 Not to be Provided	11 Method of Compensation
5.01 Project Administration	✓										E
5.02 Disciplines Coordination/Document Checking	✓										E
5.03 Agency Consulting/Review/Approval	✓										E
5.04 Owner-supplied Data Coordination	✓										E
5.21 Architectural Design/Documentation	✓										E
5.22 Structural Design/Documentation	✓										E
5.23 Mechanical Design/Documentation	–	✓ M									E
5.24 Electrical Design/Documentation	–	✓ E									E
5.25 Civil Design/Documentation	✓										E
5.26 Landscape Design/Documentation					✓						F
5.27 Interior Design/Documentation	✓										E
5.28 Materials Research/Specifications	✓										E
5.30 Special Bidding Documents/Scheduling										✓	–
5.32 Statement of Probable Construction Cost	✓										E
5.33 Presentations	✓										E
5.71 Graphics Design	✓										E

METHODS OF COMPENSATION

A = Multiple of Direct Salary Expense
B = Multiple of Direct Personnel Expense
C = Professional Fee Plus Expenses
D = Percentage of Construction Cost
E = Stipulated Sum
F = Hourly Billing Rates
G = Multiple of Amounts Billed to Architect
H = Other:

*Requires separate authorization and compensation

each such service should be developed at the time of entry and added to the descriptions for the appropriate phase(s) or included as an appendix to B162.

—In the last column, indicate the understanding between client and A/E as to method of compensation, using the appropriate identifying letter selected from the bottom of the form. The descriptions and amounts for the several methods of compensation are included in AIA Document B161, Standard Form of Agreement Between Owner and Architect for Designated Services, Article 14.

—Draw a line through all blank spaces.

The phases and their corresponding forms are as follows:

F810	All Phases	SHORT FORM
F811	Phase 1.	PREDESIGN SERVICES
F812	Phase 2.	SITE ANALYSIS SERVICES
F813	Phase 3.	SCHEMATIC DESIGN SERVICES
F814	Phase 4.	DESIGN DEVELOPMENT SERVICES
F815	Phase 5.	CONSTRUCTION DOCUMENTS SERVICES
F816	Phase 6.	BIDDING OR NEGOTIATIONS SERVICES
F817	Phase 7.	CONSTRUCTION CONTRACT ADMINISTRATION SERVICES
F818	Phase 8.	POST-CONSTRUCTION SERVICES
F819	Phase 9.	SUPPLEMENTAL SERVICES

Phase 9 contains supplemental services the A/E may be called upon to provide during any one of several of the sequential phases 1-8. The A/E may mark these on Form F819 or may prefer to enter them in an appropriate phase on Forms F811-818.

If the scope of these services is modified later, compensation should be adjusted accordingly.

Forms F810-819 may be used simply as a basis for determining costs, and as part of AIA Document B162. The forms alone are available separately as worksheets for initial entries. They are also included in B162 for entry of final decisions and signing by owner and architect. Exhibit 3-2 shows one phase of a typical preprinted *Schedule of Designated Services Worksheet* form filled in for a hypothetical project. The conditions of the project are set out in Exhibit 3-1.

4. Estimating Costs of Services and Compensation

EXHIBIT 4-1. USE GUIDELINES AT THREE LEVELS OF COMPLEXITY

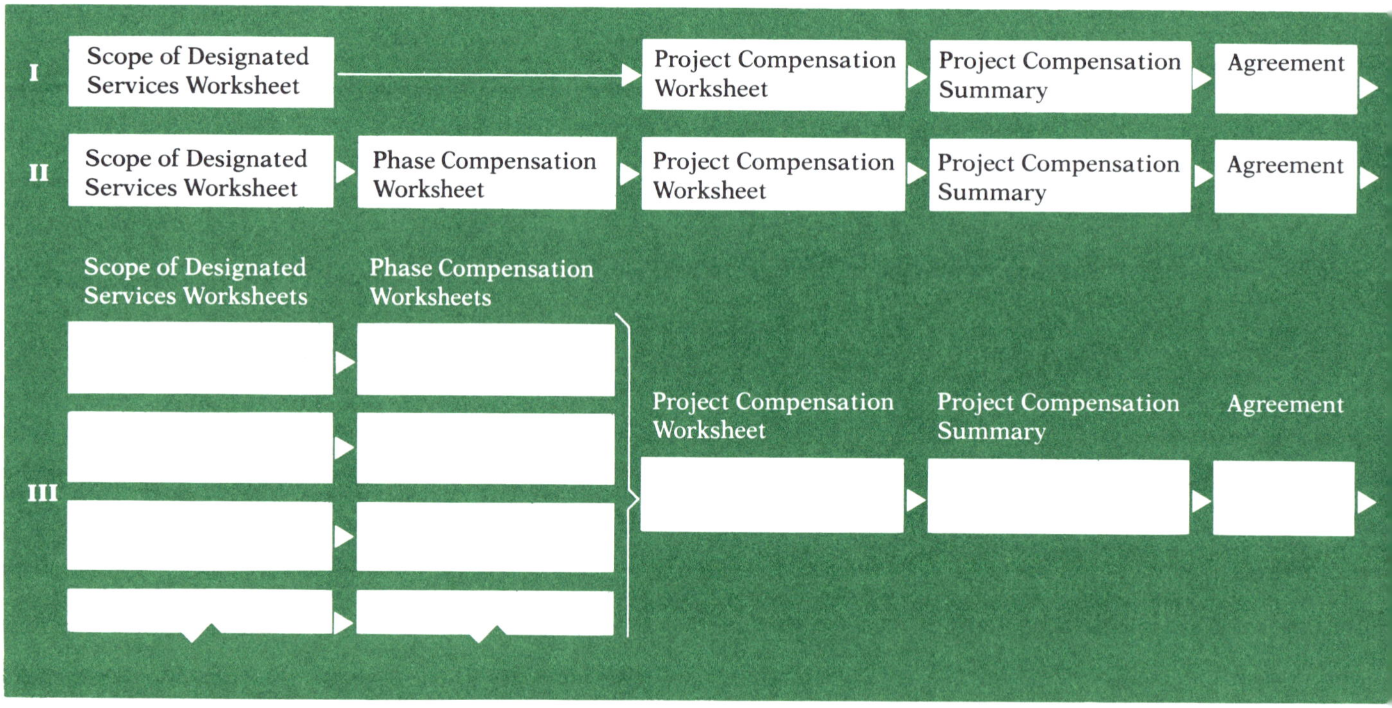

After establishing, jointly with the client, all items of service required for the project and allocating responsibility for completing these, the A/E moves on to the next step. This consists of calculating the amount of compensation that will serve as the basis for presentation to the client.

APPROACH

The A/E estimates the amounts of time and costs required to provide the services determined in the previous step, adds an amount for planned profit to arrive at proposed compensation, and estimates an amount for reimbursable expenses. To help develop and tabulate this information, AIA Forms F820-829 *(Phase Compensation Worksheet)* are available. These are normally considered in-house documents, but some A/Es may wish to share the derived data with the client when estimating compensation. (This effort should be only as detailed as required by the size and scope of the project. Several simplified methods are discussed later in this chapter.)

The forms have been developed for maximum flexibility. They may be applied to large or small, complex or simple projects, and with detailed or more generalized breakdowns of estimated time and cost. They are geared for use by both large and small firms, those with sophisticated accounting systems and those with more simplified bookkeeping procedures (Exhibit 4-1).

One copy of forms F820-829 is included in Appendix D of the *Guidelines*. These are part of the overall series developed for the AIA Financial Management System. Forms F821-829 come *with* designated services listed by phase, with spaces for additional services; form F820 comes *without* the listing. This a lows a firm to tailor services listings t the needs of a specific project or to th firm's operations. The firm may desig its own forms to suit its own needs an procedures in applying the cost-base concepts.

Aside from their use for calculatin proposed compensation and estimate reimbursable expense, these forms als are a tool for improving a firm's busines practices and upgrading its time ac counting procedures, since they assum detailed knowledge of the time and cost required for the various items of service

A good financial accounting syster will bring out much of the detail neede to calculate the *cost* factors. Otherwis consult other components of the AIA F nancial Management System, or an ac countant.

As for *time*, estimating the actua working time required to perforr

EXHIBIT 4-2. CASE STUDY OF AN A/E FIRM

The hypothetical architectural/engineering firm of Apple and Bartlett Associates has a total staff of 12, broken down at the following staff classifications with hourly salary expense based on a 2080-hour gross work year.

Staff Classification*	Annual	Hourly
1. PRINCIPAL ("salary" or "draw" only)		
Apple	$ 29,120	$14.00
Bartlett	29,120	14.00
2. SUPERVISORY		
Charles	20,800	10.00
Durango	20,800	10.00
3. TECHNICAL I		
Edward	16,640	8.00
Frances	16,640	8.00
4. TECHNICAL II		
Ginger	12,480	6.00
Harold	12,480	6.00
Irving	12,480	6.00
5. TECHNICAL III		
Joan	8,320	4.00
Karl	8,320	4.00
Lewis	8,320	4.00
Total	$195,520	

*See Appendix A

Total paid time off of 30 days per year (240 hours) per person equals $22,560, leaving $172,960 for working time salary expense.

The cost of benefits (mandatory and customary, including paid time off) has averaged 25% of working time salary expense for the past twelve months.

For the purpose of this example the firm's indirect expense is 84% of (working time) direct personnel expense, or 130% of (working time) direct salary expense (see Exhibit 2-5).

services requires a clear understanding of the tasks that make up the service. This is by and large a matter of judgment and experience, supported by reliable in-house time and cost records from past projects. Proposed for development by the AIA is a time data bank which will enable practitioners to compare phase time data on similar projects. Its data base should be most useful for checking personnel time requirements experienced on other firms' projects against those developed from in-house data.

After the A/E has completed the Phase Compensation Worksheet, a Project Compensation Worksheet (Form F830) is provided for summarizing the estimates to date, chiefly for internal use. Following that, the practitioner completes a Project Compensation Summary (F840) for discussions with the client.

The suggested steps in completing these forms are described below. (Before completing the forms, it is a good idea to examine again Chapter 2, *Compensation Principles and Definitions.*)

To further illustrate these steps, a case study has been prepared based on the compensation structure of the hypothetical A/E firm. The office structure, shown in Exhibit 4-2, is imaginary as to description, books of the firm and the reality of the assumed figures.

HOW TO COMPLETE THE PHASE COMPENSATION WORKSHEET (F820)

Regular "Long Form" Method. The following procedure is recommended in filling out the *Phase Compensation Worksheet* form (Exhibit 4-3). A typical phase (Construction Documents) has been selected.

Step 1. When using the blank form F820, enter along the top those services to be provided or coordinated by the A/E firm. On preprinted forms, cross out services that do not apply, and add others as needed. If services have been added to the preprinted *Schedule of Designated Services Worksheet* form, they should be similarly added to the preprinted *Phase Compensation Worksheet* form. If the blank *Schedule of Designated Services Worksheet* form is used, the blank *Phase Compensation Worksheet* form (F820) should conform with matching entries.

Step 2. Enter the estimate of the total number of direct in-house working hours required to perform a given service (line 6, upper boxes). When the column is for an outside service (as in items 5.23 and 5.24), obtain this figure from the consultant for record purposes and enter the amount in the horizontal line marked "Outside Services Expense" (line 11).

Step 3. Break down hours for each service into personnel levels. The form provides horizontal lines to accommodate five established levels of personnel (see Appendix A); this is consistent with other components of the AIA Financial

EXHIBIT 4-3. SAMPLE COMPLETED PHASE COMPENSATION WORKSHEET FORM F825

PHASE COMPENSATION WORKSHEET

PHASE 5: CONSTRUCTION DOCUMENTS SERVICES

Project: Franklin Elementary

Project #: 7701 Date: 1.10.77

Owner: Alton School Board

Architect: Apple & Bartlett Assoc.

SERVICE		5.01 Project Administration	5.02 Disciplines Coord./ Document Checking	5.03 Agency Consulting/ Review/Approval	5.04 Owner-supplied Data Coordination	5.21 Architectural Design/ Documentation	5.22 Structural Design/ Documentation	5.23 Mechanical Design/ Documentation	5.24 Electrical Design/ Documentation	5.25 Civil Design/ Documentation	5.26 Landscape Design/ Documentation	5.27 Interior Design/ Documentation	5.28 Materials Research/ Specifications	~~5.30 Special Bidding Documents/Scheduling~~	5.32 Statement of Probable Construction Cost	5.33 Presentations	5.71 Graphics Design
IN-HOUSE PERSONNEL (line 1)	Hrs.	90	60	20	4	60	40			20	4	14	40			8	
	$	1,260	840	280	56	840	560			280	56	196	560			112	
(line 2)	Hrs.	12	20	10	12	80	120			40	8		80		36	8	40
	$	120	200	100	120	800	1,200			400	80		800		360	80	400
(line 3)	Hrs.		20			140	80					50					20
	$		160			1,120	640					400					160
(line 4)	Hrs.	10	6			240				40							
	$	60	36			1,440				240							
(line 5)	Hrs.	20				40						20	70		8		
	$	80				160						80	280		32		
SUB TOT.	Hrs.	132	106	30	16	560	240			100	12	84	190		44	16	60
	$	1,520	1,236	380	176	4,360	2,400			920	136	676	1,640		392	192	560
OUT-SIDE	Hrs.																
	$							11,200 (5.23–5.24)									
TOT.	Hrs.																
	$																

TOTAL HOURS	TOTAL DOLLARS	ITEM	LINE
360	5,040	Principal @ $ 14	1
466	4,660	Supervisory @ $ 10	2
310	2,480	Tech. I @ $ 8	3
296	1,776	Tech. II @ $ 6	4
158	632	Tech. III @ $ 4	5
1,590	14,588	Direct In-house Salary Expense	6
	—	Direct Personnel Expense	7
	18,964	Indirect Expense @130% of 6	8
	1,007	Other Nonreimbursable Direct Expense @3% of 6+8	9
	34,559	Total In-house Expense 6+8+9	10
	11,200	Outside Services Expense	11
	45,759	Estimated Total Expense 10+11	12
	3,400	Contingency	13
	7,300	Profit	14
	56,459	Proposed Compensation 12+13+14	15
	750	Estimated Reimbursable Expense	16

REMARKS

ALTERNATE COMPUTATION

	ALTERNATE @ $	
	COMPUTATION @ $	5
14,588	Direct In-house Salary Expense	6
18,235	Direct Personnel Expense @125% of 6	7
15,317	Indirect Expense @84% of 7	8
1,007	Other Nonreimbursable Direct Expense @3% of 7+8	9
34,559	Total In-house Expense 7+8+9	10
	Outside Services Expense	11
	Estimated Total Expense	

AIA® FORM F825 • PHASE COMPENSATION WORKSHEET, PHASE 5 • JANUARY 1978 EDITION • © 1978

THE AMERICAN INSTITUTE OF ARCHITECTS, 1735 NEW YORK AVE., N.W., WASHINGTON, DC 20006

THE AMERICAN INSTITUTE OF ARCHITECTS FINANCIAL MANAGEMENT SYSTEM

Management System.

(Certain clients and government agencies may require this type of breakdown. Should the project require overtime work, allocation between regular hours and overtime hours should also be noted on a separate sheet.)

Step 4. Enter the salary rate for each level in the box at the right end of each line (lines 1-5). Standardized or average hourly rates may be used for each of the several levels. For overtime work, the premium should be factored into the hourly rate as appropriate. For long-term projects, anticipate salary increases.

Step 5. Service boxes in lines 1-5 are divided horizontally. Enter hours above the dividing line (upper boxes). Multiply by the rate and enter the direct salary amount below the dividing line (lines 1-5, lower boxes).

Step 6. Subtotal (vertically) direct in-house salary amounts (line 6, lower boxes under services). Total hours and dollar amounts for salaries (from upper and lower boxes under services in lines 1-6) horizontally, in appropriate columns at right under "total hours" and "total dollars," respectively (lines 1-6).

Step 7. If the firm uses direct personnel expense, enter an amount for this item by applying the firm's benefits factor as defined in Chapter 2 (line 7). (See alternate computation, Exhibit 4-3.)

Step 8. Multiply the amount on line 6 or 7 by the appropriate established indirect expense factor as defined in Chapter 2. This may be a ratio or percentage. Indicate if the factor is applied to direct salary or direct personnel expense (line 8). Note that direct personnel benefits should be added to indirect expense if line 6 is used as the base; if line 7 is used, they should not be, since they are already considered as part of direct personnel expense.

Step 9. Enter the total of other nonreimbursable direct expense (line 9). To describe in more detail, the Remarks box may be used.

Step 10. If direct personnel expense is used, total the amounts in lines 7-9 to obtain total in-house expense (line 10). If DPE is not used, total lines 6, 8 and 9.

Step 11. Enter outside services expense (line 11)—in this case the services of mechanical and electrical engineering consultants—and add to line 10 to obtain estimated total expense (line 12).

(Enter *actual* quotations or *estimated* costs of professional consultants. Also enter estimated hours if available.)

Step 12. Enter an amount to allow for project contingencies (line 13). The amount entered is a matter of judgment; it considers possible project variables as discussed in Chapter 2. This line may be left blank if it is judged that adequate contingencies have been included in time budgeting.

Step 13. Enter an amount for planned profit (line 14). (Review the discussion in Chapter 2.) The entry in this example is hypothetical and not a recommendation.

Step 14. Total lines 12-14 to obtain proposed compensation (line 15).

Step 15. Determine estimated reimbursable expenses. List the items and their estimated costs on a backup sheet (or in the Remarks box if few in number). Enter the total in line 16. The client is charged for these expenses in addition to compensation. If the A/E proposes a markup on these items, this should be noted in the calculations and taken into account when determining planned profit.

"Short Form" Alternatives.
Following is a selection of simplified methods for completing the *Phase Compensation Worksheet* forms, for use with small or simple projects:

Alternative A.

Step 1. Estimate the total amount of in-house direct salaries or direct personnel expense for the *total* project or by complete phases.

Step 2. Apply an appropriate multiplier to this amount to cover indirect and other direct expenses and planned profit.

Step 3. Add an estimate of the costs of outside services (with markup) and a contingency factor if appropriate.

Step 4. Total Step 2 and Step 3 results to obtain total estimated compensation.

Step 5. Estimate reimbursable expenses separately from compensation.

Alternative B.

Step 1. Lump items of service within each phase into related categories.

Step 2. Estimate the hours by category.

Step 3. Convert them to estimated total compensation as in Alternative A.

Alternative C.

Step 1. Lump related items of service within a phase or use individual items.

Step 2. Estimate direct hours for each.

Step 3. Multiply by a rate to produce direct salaries or direct personnel expense.

Step 4. Total these amounts for the phase.

Step 5. Apply a factor for their share of indirect expense.

Step 6. Add nonreimbursable direct expense; this produces in-house expense for the phase.

Step 7. The consultant firm calculates its services in the same way; the total is added to in-house expense.

Step 8. Add up the results and apply factors for profit; this gives proposed compensation for the phase.

Step 9. Estimate reimbursable expense for the phase.

Alternative D.

Step 1. Analyze the number and type of drawings in terms of direct salaries or

EXHIBIT 4-4. SAMPLE COMPLETED PROJECT COMPENSATION WORKSHEET FORM F830

PROJECT COMPENSATION WORKSHEET

Project: *Franklin Elementary*

Project #: *7701* Date: *1.10.77*

Owner: *Alton School Board*

Architect: *Apple & Bartlett Assoc.*

PHASE				3 *Schematic Design*	4 *Design Development*	5 *Construction Documents*	6 *Bidding or Negotiations*	7 *Construction Contract Administration*				TOTAL HOURS	TOTAL DOLLARS	ITEM	LINE
IN-HOUSE PERSONNEL	Hrs.					360								@ $	1
	$					5,040									
	Hrs.					466								@ $	2
	$					4,660									
	Hrs.					310								@ $	3
	$					2,480									
	Hrs.					296								@ $	4
	$					1,776									
	Hrs.					158								@ $	5
	$					632									
SUB TOT.	Hrs.			370	647	1,590	142	770				*3,519*	*32,417*	Direct In-house Salary Expense	6
	$			3,890	5,835	14,588	1,297	6,807							
	$			4,862	7,294	18,235	1,621	8,509					*40,521*	Direct Personnel Expense *@125% of 6*	7
	$			4,084	6,127	15,317	1,362	7,148					*34,038*	Indirect Expense *@84% of 7*	8
	$			268	403	1,007	89	470					*2,237*	Other Nonreimbursable Direct Expense	9
	$			9,214	13,824	34,559	3,072	16,127					*76,796*	Total In-house Expense	10
OUT-SIDE	Hrs.												*25,200*	Outside Services Expense	11
	$			1,500	6,000	11,200	750	5,750							
TOT.	Hrs.												*101,996*	Estimated Total Expense	12
	$			10,714	19,824	45,759	3,822	21,877							
	$												*6,854*	Contingency	13
	$												*16,150*	Profit	14
	$												*125,000*	Proposed Compensation	15
	$												*5,900*	Estimated Reimbursable Expense	16

THE AMERICAN INSTITUTE OF ARCHITECTS
FINANCIAL MANAGEMENT SYSTEM

irect personnel expense.

Step 2. Total the amounts.

Step 3. Proceed as in Alternative C, tep 5.

Alternative A may be recorded diectly on Form F830 *(Project Compensaon Worksheet)* as discussed below.

The value of simplified or abbreviated nethods depends upon determining appropriate multipliers and factors. These end to be more reliable when derived rom each firm's own experience. nalyzing the firm's books and records vill provide data from which such multipliers and factors may be derived. On he other hand, detailed and more ophisticated cost analysis techniques han are discussed in these *Guidelines* nay be used to produce data that need nly be *summarized* on the individual hase worksheets.

After estimating the amounts of comensation and reimbursables for the diferent phases of a project, summarize he detailed data for in-house review, reast the data as needed for review with he client, and discuss with the client the mount and method of compensation. orms F830 and F840 have been develped for this purpose.

IOW TO COMPLETE THE 'ROJECT COMPENSATION VORKSHEET (F830)

'ime and cost figures from the "totals" olumn of each *Phase Compensation Vorksheet* are entered under the approriate phase column of the *Project Comensation Worksheet* form (Exhibit 4-4). Iorizontally add the amounts on lines 6 r 7 thru 16 and enter in the "totals" olumn at the right. The sum of these otals is the proposed compensation for he project and total estimated reimursable expense.

If no services are to be provided in a pecific phase, do not enter the phase title, and leave that column blank on the *Project Compensation Worksheet.*

If the A/E wishes to share detailed derived figures when discussing compensation with the client, the *Project Compensation Worksheet* (F830) and *Summary* (F840) forms are used jointly with the *Schedule of Designated Services Worksheet* form (F810-F819); if not, F830 should be considered only as an in-house document together with the F820 series. Entries in these forms may require revisions at this time, as may the description or scope of certain items of service.

For Simple Projects.

Besides serving as a summary or recapitulation of time and cost data developed on the phase worksheet (F820 series), the project worksheet is also designed to be used in lieu of the phase worksheet for simpler projects that do not require the degree of breakdown of time and cost estimates. See also description of short-form alternatives earlier.

An example has been worked out to demonstrate the process (see Exhibit 4-5). Our A/E firm, Apple & Bartlett Associates, was asked to investigate roofing problems at Jefferson High School, prepare a report and contract documents, and administer the contract to repair the leaks. The following steps are recommended when using Form F830 for this purpose.

Step 1. Enter in the appropriate blocks on the top line the phases or specific activities required by the project.

Step 2. Enter the appropriate time and cost estimates for each salary level to be assigned to the project.

Step 3. Total these amounts at the right.

Step 4. Henceforth proceed as though completing the phase compensation worksheet, steps 7 to 15 detailed on page 15.

HOW TO COMPLETE THE PROJECT COMPENSATION SUMMARY FORM (F840)

The Project Compensation Summary form (Exhibit 4-6) may be used to enter summarized data from the Project Compensation Worksheet for discussion with the client. The following steps are recommended:

Step 1. In the "Phase" column, enter the names of phases or other groupings of services.

Step 2. In the column marked "Proposed Compensation Amount by Phase," enter appropriate totals from the project worksheet. If no services are to be provided in a specific entered phase, draw a line through that space.

Step 3. The next column is used for developing cumulative totals of compensation. Add progressively the amounts in the preceding column and enter the totals here. A simpler alternative procedure is to enter only the cumulative totals, without completing the individual phase line items. The last entry is the total estimated compensation.

Step 4. In the column marked "Estimated Reimbursable Expense," enter the appropriate amounts from the project worksheet.

Step 5. The following two columns may be used to enter summary data resulting from discussions between the client and A/E concerning revisions to the proposed scope and compensation. The blank last two columns may be used for further revisions or for adding new relevant data.

HOW TO COMPLETE THE TIME AND PAYMENT SCHEDULE (F850)

The *Project Time and Payment Schedule* Form (AIA Form 850) completes the family of forms developed for use with the *Guidelines* (other than the owner-

EXHIBIT 4-5. EXAMPLE OF PROJECT COMPENSATION WORKSHEET FORM F830 USED FOR A SIMPLE PROJECT

PROJECT COMPENSATION WORKSHEET

Project: Roof Repairs
Project #: 7622 Date: 10·10·76
Owner: Jefferson High School
Architect: Apple & Bartlett Assoc.

PHASE		Project Administration	Investigation & Research	Report	Construction Documents	Bidding or Negotiations	Construction Contract Administration					TOTAL HOURS	TOTAL DOLLARS	ITEM	LINE
IN-HOUSE PERSONNEL	Hrs.	8	4	2	4	4						22	308	Principal @ $ 14	1
	$	112	56	28	56	56									
	Hrs.		16	16	40		20					92	920	Supervisory @ $ 10	2
	$		160	160	400		200								
	Hrs.													Tech. I @ $ 8	3
	$														
	Hrs.													Tech. II @ $ 6	4
	$														
	Hrs.			6	20							26	104	Tech. III @ $ 4	5
	$			24	80										
SUB TOT.	Hrs.											140	1,332	Direct In-house Salary Expense	6
	$														
	$												—	Direct Personnel Expense	7
	$												1,732	Indirect Expense @ 130% of 6	8
	$												90	Other Nonreimbursable Direct Expense	9
	$												3,154	Total In-house Expense	10
OUT-SIDE	Hrs.												—	Outside Services Expense	11
	$														
TOT.	Hrs.												3,154	Estimated Total Expense	12
	$														
	$												300	Contingency	13
	$												800	Profit	14
	$												4,254	Proposed Compensation	15
	$												150	Estimated Reimbursable Expense	16

AIA® FORM F830 • PROJECT COMPENSATION WORKSHEET • JANUARY 1978 EDITION • © 1978
THE AMERICAN INSTITUTE OF ARCHITECTS, 1735 NEW YORK AVE., N.W., WASHINGTON, DC 20006

THE AMERICAN INSTITUTE OF ARCHITECTS
FINANCIAL MANAGEMENT SYSTEM

EXHIBIT 4-6. SAMPLE COMPLETED PROJECT COMPENSATION SUMMARY FORM F840

PROJECT COMPENSATION SUMMARY

Project Franklin Elementary
Project # 7701 Date 1·10·77
Owner Alton School Board
Architect Apple & Bartlett Assoc.

PHASE	Proposed Compensation Amount By Phase	Cumulative Proposed Compensation Amount	Estimated Reimbursable Expense	Date	Revision		
Schematic Design	13,020	13,020	2,000				
Design Development	24,180	37,200	150				
Construction Documents	56,459	93,659	750				
Bidding or Negotiations	4,588	98,247	1,500				
Construction Contract Administration	26,753	125,000	1,500				
TOTALS		125,000	5,900				

REMARKS

AIA® FORM F840 • PROJECT COMPENSATION SUMMARY • JANUARY 1978 EDITION • © 1978
THE AMERICAN INSTITUTE OF ARCHITECTS, 1735 NEW YORK AVE., N.W., WASHINGTON, DC 20006

THE AMERICAN INSTITUTE OF ARCHITECTS
FINANCIAL MANAGEMENT SYSTEM

EXHIBIT 4-7. SAMPLE COMPLETED PROJECT TIME AND PAYMENT SCHEDULE FORM F850

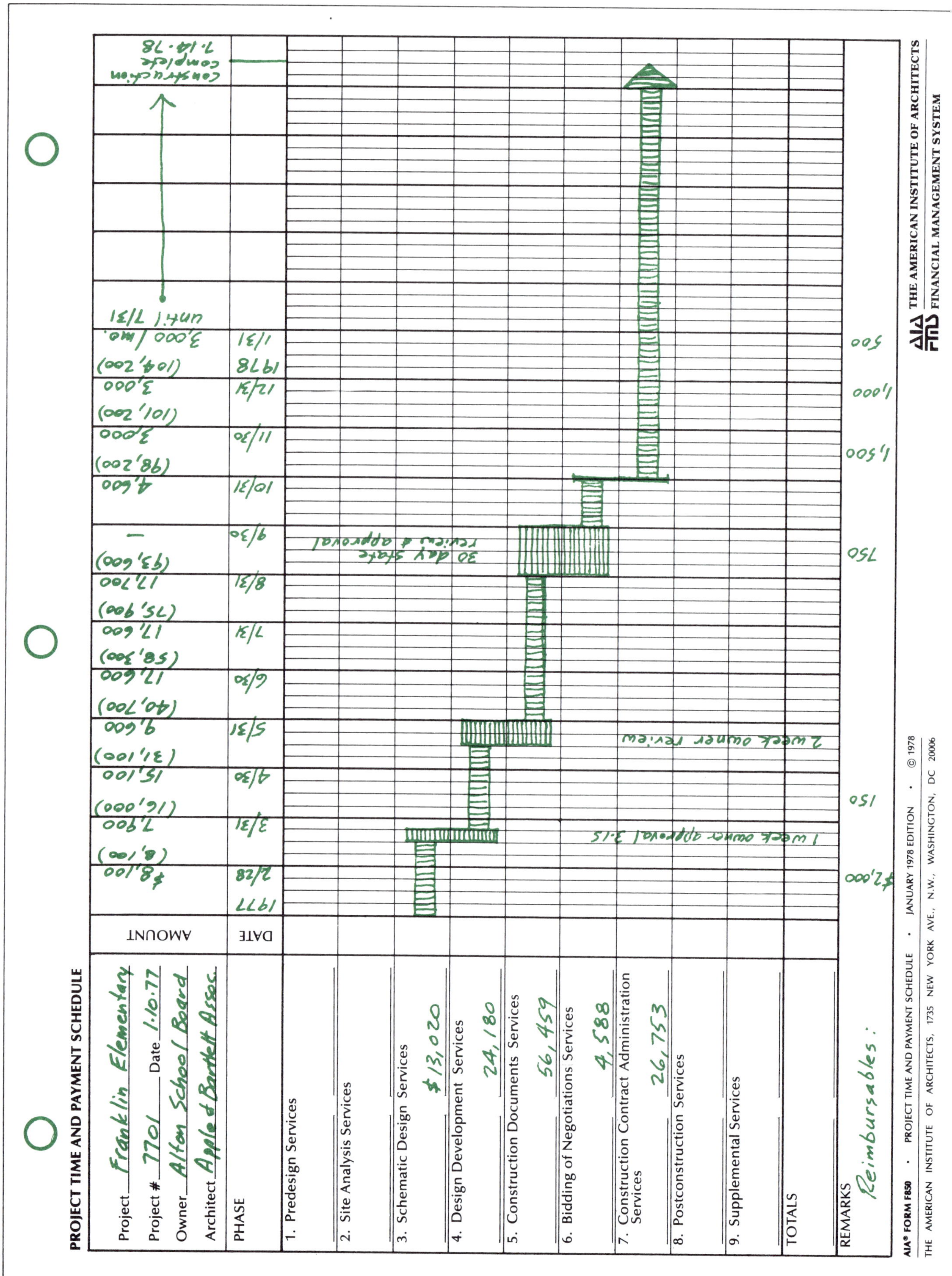

chitect agreement forms). While optional, the form is valuable for estimating the time frame in which the agreed-upon services are to be rendered and payments made (Exhibit 4-7).

The *Project Time and Payment Schedule* form represents, graphically, the estimated elapsed time involved in performing certain phases of services, and the breakdown of scheduled payments the client to the A/E. This form can become a time and personnel planning guide for the A/E and a cash flow guide for both the client and A/E. It is suggested that this form not be made part of any agreement between owner and architect.

The recommended steps in completing this form are:

Step 1. Enter the estimated time frame in which the work is to be completed across the chart in the line labelled "Date." The left-to-right graduations can represent days, weeks or months to reflect the time span of services. In the example, each block represents one month.

Step 2. Obtain from the phase and project compensation worksheets the required working time and compensation figures for each phase, convert working time to elapsed time, and enter estimated elapsed time frame data on the appropriate horizontal line. Use a solid bar running from the estimated starting date to the estimated completion date for the work in each phase. (The time frames for the phases may be sequential or overlapping, depending on the project's needs.)

Step 3. Indicate payment due dates by vertical lines drawn through the chart on the dates payments are due. It is recommended that these be periodic (monthly, semi-monthly, etc.) rather than only at the end of each phase.

Step 4. Enter estimated amounts of payments for each period over the payment line in the column titled "Amount." At each date of payment, that portion (percentage) of the total phase work completed since the last date of payment is considered billable work. The dollar amounts payable may be divided equally among the periods representative of the total compensation established for that phase. Alternatively, if the workload is expected to vary from period to period, dollar amounts should be made proportional to the amount of expected effort. On the other hand, if payments are to be made for actual time expended to date, the form would be used simply to monitor estimated time and charges.

Step 5. Under each payment entry, also enter, in parentheses, the cumulative amount of compensation to that date.

Step 6. (optional) Note that blank space appears under each phase title; this may be used to enter the total hours and amount of compensation allocable to that phase.

Step 7. Enter reimbursables in the "Remarks" box.

As the client and the A/E discuss the scope, description of services, compensation and time frame for the project, the method of compensation enters strongly into the picture, even though the *Guidelines* themselves do not concern themselves with this aspect.

All methods of compensation take into account the elements of expense and profit that together constitute compensation, plus the items of reimbursable expense. The methods include:

A. Multiple of direct salary expense
B. Multiple of direct personnel expense
C. Professional fee plus expenses
D. Percentage of construction cost
E. Stipulated sum
F. Hourly billing rates
G. Multiple of amounts billed to architect
H. Other
   1. Square footage
   2. Unit cost
   3. Royalty
   4. Combined methods

Refer to Appendix B for a detailed discussion of the various methods.

The determination of specific compensation for services agreed upon between client and architect/engineer is influenced by:

—The A/E's experience in estimating personnel-hours and dollars for a specific effort, or access to reliable sources of information;

—The existence of a well-defined program for the project;

—The client's experience in dealing with a comprehensive approach to compensation and the presence (or absence) of a client's predetermined "fee curve" or other set system;

—The project's type and scope;

—The A/E's and client's readiness to negotiate with regard to scope and value of effort;

—The role and involvement of the A/E's consultants.

Of these, the first two items tend to have the major impact in determining compensation. If the A/E includes programming as a service, preferably prior to pinning down the remaining services, the knowledge gained thereby can be helpful in sharpening time estimates for the project.

5. Agreements for Professional Services

Earlier chapters discussed methods for determining the scope of professional services, estimated costs and compensation. This chapter deals with the final step, the working up of terms for the formal agreement between owner and architect or engineer in which scope and compensation are set out in detail.

INTRODUCTION

The architectural and engineering societies have produced various standard forms of agreement to make the legal aspects of contracting for services as straightforward and error-free as possible, as well as to protect public, corporate or private signatories. The forms are designed to be used in conjunction with other documents developed by these societies, such as the agreements between owner and contractor or between architect and engineer, and the general conditions of the contract for construction.

These documents are the product of the combined efforts of the professional societies, owner and contractor representatives, and legal and insurance counsels. Revisions, reflecting the latest thinking of experts in many disciplines, are made to the documents at reasonable intervals. The forms of agreement allow the A/E to enter into a number of compensation arrangements when used with the *Guidelines*.

As pointed out in the section on agreements in the 1977 edition of the American Consulting Engineers Council's *Manual of Practice*, standard forms of agreement and other forms for the guidance of engineers are endorsed by ACEC and the Professional Engineers in Private Practice division of the National Society of Professional Engineers. One referenced form is the 1975 Standard Form of Agreement between Owner and Engineer for Professional Services, Document #4, obtainable from ACEC headquarters in Washington, DC. It offers a choice of three different methods of compensation and provides for describing basic and additional services.

The A/E and client are not required to use AIA or ACEC documents. Any form of agreement which is equitable, and upon which the client and A/E concur, is appropriate. If a non-AIA or non-ACEC form of agreement is to be used in conjunction with the *Guidelines*, it should be evaluated by legal counsel, as should any modifications to standard AIA and ACEC documents (see also Ref. 1, Appendix C for an article dealing with use of non-AIA forms).

AIA FORMS OF AGREEMENT

The latest edition of Chapter 9, "Owner-Architect Agreements," of the AIA's *Architect's Handbook of Professional Practice* discusses general principles and the currently available series of AIA standard forms of agreement between owner and architect. Identified below with their abbreviated titles, these include:

B141	Owner-Architect Agreement
B141/CM	Owner-Architect Agreement, Construction Management Edition
B151	Abbreviated Owner-Architect Agreement
B161	Designated Services Agreement
B162	Scope of Designated Services
B171	Interior Design Services Agreement
B727	Special Services Agreement

AIA Document B141, the standard the profession for many years, provid in its latest version an unlimited choi of methods of compensation for bo basic and additional services. Basic se vices consist of defined architectur services and normal related structur mechanical and electrical engineerin services for Phases 3-7. All others a considered additional services.

AIA Documents B161 and B162 a new and have been developed speci cally for use with the *Guidelines*. Th "boilerplate" which evolved over th years for B141 has been adapted to B16 thus maintaining the standards an fairness which users have come to expe of these documents. In combinatio with the specific services designated B162, the B161 owner-architect agre ment form provides a choice or choic of seven defined methods of basic com pensation, plus unlimited addition methods. Space is also provided fo specifying compensation for servic which may be required by the own later in addition to those originally de ignated.

Examples of both B161 and B162 a shown in Appendix D, together wi their related instruction sheets. Study these exhibits will aid in understandin application of the *Guidelines*.

5. Monitoring Compensation

he *Guidelines* can and, indeed, should e used for more than estimating and egotiating compensation. This chapter oints up additional benefits of the ap- roach.

'OOL FOR INTERNAL FFICE MANAGEMENT

t the stage where the agreement for rofessional services is drawn up, with ll the preparatory work this requires, he architect/engineer has in hand a val- able tool for internal office manage- nent that can benefit both the A/E firm nd the client. A schedule of designated ervices, an estimated cost for each of hose services, and usually a time sched- le for completing each service, have een established. If the estimate is not ivided into parts by each service, it is robably broken down by groups of ervices or by phases, but the manage- nent potentials are the same.

If the A/E uses this information to udget and monitor hours and dollars, hese benefits can be expected:

—*A communication benefit.* The staff ecomes aware of schedules and expec- ations, with early opportunity for feed- ack of information. The client gets to now how much will be billed (and vhen), establishing a project cash flow rojection.

—*An accountability benefit.* The ap- roach provides goals (in hours and dol- ars) for each specific service, and the neans of measuring progress towards hese goals. If those individuals respon- ible for meeting goals had a part in the stimating, so much the better. The ap- roach could stimulate greater effi- iency and productivity, with resulting enefits to A/E and client.

—*Future estimating benefits.* By estab- ishing an office data bank of time and osts linked to specific services and nown project conditions, the A/E is rovided with a much more reliable basis for future estimating of time and costs.

The budget and monitoring can take several forms. Depending on the detail of the original estimate and the sophistication of the available office time and cost information system, the budgeting may be by *hours only, cost only,* or *both,* and may be established on the basis of *disciplines, phases, services* and *personnel classification,* or simply by *total project,* or by any level of detail between these extremes. The degree of detail is as reliable as the accumulated data, but not more so.

Budgeting may or may not parallel estimating, depending on the size and complexity of the project, office philosophy and the quality of the firm's time and cost management system. It helps to use the *Schedule of Designated Services* at the estimating stage to assist in avoiding oversights and omissions. To budget and monitor in similar detail may not be desirable, necessary or possible on many projects. To go a step beyond budgeting by phases only, related services might be grouped for timekeeping purposes under a limited number of headings per phase. (For further discussion of budgeting and scheduling, see Appendix C, Reference 5, Chapter 10.)

TIMEKEEPING AND THE AIA COMPUTER-BASED FINANCIAL MANAGEMENT SERVICE

The AIA Computer-based Financial Management Service (CFMS) is capable of budgeting and monitoring all services presently listed in the *Scope of Designated Services* (Appendix A and Appendix D), plus a limited number of additional services. It also permits budgeting and tracking by office staff classifications of personnel for each service and each phase, and by office departments or disciplines.

For monitoring, record project time consistent with the budgeted entries. Then, for each reporting period, add up the hours, multiply by the appropriate direct salary or direct personnel rates and add total hours and dollars for each entry heading. Using both *identifying numbers* and *descriptive headings* for each phase and service on office time records cuts the risk of entry errors and makes for easier supervisory checking; but for entries into a manual system or CFMS, only the identifying numbers are normally necessary.

The timekeeping operation may be done manually or by using CFMS at several levels of complexity. Manual calculation of hours and direct salary expense (or direct personnel expense) is feasible but may become unwieldy and time-consuming at a certain level of reporting.

Most firms start with the *basic level* of CFMS, which is its lowest cost module. This module allows a firm to compare estimated project hours and dollars for a reporting period against actual costs on a time basis. Further, if estimated "percent complete" figures for the individual services are entered, CFMS displays the prorated portions of the budgeted hours and dollars, based on these estimated "percent completes." It also displays a comparison of the *estimated* and *calculated* "percent completes." The printout (Exhibit 6-1) illustrates the format of the CFMS project progress report for comparing project costs and budgets.

After a firm has become familiar with the basic level of CFMS, it may choose to upgrade via additional modules to the *intermediate level* of CFMS. This intermediate level offers the ability to budget and monitor all direct costs, indirect costs, costs for outside services and reimbursables for each project for each reporting period. Thus, the intermediate level provides for reporting all components of project cost—reimbursables, personnel time and other direct costs,

EXHIBIT 6-1. TYPICAL CFMS PROJECT PROGRESS REPORT

ANDERSON BABCOCK AND CROWLEY
ARCHITECTS/PLANNERS

PAGE 1
RUN DATE 02/06/77

PROJECT CITY HALL
NUMBER 1005.00

PROJECT PROGRESS REPORT
FOR THE PERIOD 01/01/77-01/31/77

PRINCIPAL SMITH
PROJ MGR JONES

	SPENT THIS PERIOD HOURS	SPENT THIS PERIOD DOLLARS	SPENT TO DATE HOURS	SPENT TO DATE DOLLARS	PERCENT EXPENDED	COMPLETE REPORTED	EARNED HOURS	BUDGET DOLLARS	TOTAL HOURS	BUDGET DOLLARS
ARCHITECTURAL DEPARTMENT										
PRE-DESIGN										
PROJECT ADMINISTRATION										
A1011 PRINCIPAL	10	150	30	450	75	80	32	480	40	600
A1012 SUPERVISORY	4	40	10	100	100	90	9	90	10	100
A1013 TECHNICAL I	1	8	2	16						
TOTAL PROJ ADMIN	15	198	42	566	81	81	41	570	50	700
PROGRAMMING										
A1051 PRINCIPAL	1	15	3	45	56	70	4	56	5	80
A1052 SUPERVISORY	1	10	2	20	40	50	3	25	5	50
TOTAL PROGRAMMING	2	25	5	65	50	62	7	81	10	130
TOTAL PRE-DESIGN	17	223	47	631	76	78	48	651	60	830
SITE ANALYSIS										
PROJECT ADMINISTRATION										
A2011 PRINCIPAL	3	45	10	150	38	40	12	160	30	400
TOTAL PROJ ADMIN	3	45	10	150	38	40	12	160	30	400
SITE DEVELOPMENT PLAN										
A2143 TECHNICAL I	1	8	6	48	8	5	4	30	80	600
TOTAL SITE DEVEL	1	8	6	48	8	5	4	30	80	600
TOTAL SITE ANALYSIS	4	53	16	198	20	19	16	190	110	1000
SCHEMATIC DESIGN										
PROJECT ADMINISTRATION										
A3011 PRINCIPAL	2	30	10	150	13	25	25	300	100	1200
A3013 TECHNICAL I	2	20	20	160						
TOTAL PROJ ADMIN	4	50	30	310	26	25	25	300	100	1200
ARCHITECTURAL DESIGN										
A3213 TECHNICAL I	18	144	40	320	16	10	25	200	250	2000
TOTAL ARCH DESIGN	18	144	40	320	16	10	25	200	250	2000
TOTAL SCHEMATIC DESIGN	22	194	70	630	20	16	50	500	350	3200
TOTAL ARCHITECTURAL DEPT.	43	470	133	1459	29	27	114	1341	520	5030
STRUCTURAL DEPARTMENT										
PRE-DESIGN										
EXISTING FACILITIES SURVEY										
S1071 PRINCIPAL	2	30	5	75	19	65	20	260	30	400
S1073 TECHNICAL I	1	8	10	80						
S1074 TECHNICAL II	1	6	20	120						
TOTAL EXIST FAC	4	44	35	275						
TOTAL PRE-DESIGN	4	44	35	275	69	65	20	260	30	400
SCHEMATIC DESIGN										
STRUCTURAL DESIGN										
S3221 PRINCIPAL	2	30	8	120	8	5	10	80	200	1600
TOTAL STRUCTURAL DESIGN	2	30	8	120	8	5	10	80	200	1600
TOTAL SCHEMATIC DESIGN	2	30	8	120	8	5	10	80	200	1600
TOTAL STRUCTURAL DEPT.	6	74	43	395	20	17	30	340	230	2000
TOTAL LABOR	49	544	176	1854	26	24	144	1681	750	7030

EXHIBIT 6-2. EXAMPLE OF A TIME RECORD FORM

TIME RECORD—AIA FORM • F301

Project #	Sub	Department	Phase	Service	Staff Level	Employee Number	Payroll Ending Date	Name
							/ /	
						Project Name		Day/Date—Hours Worked

Total Hours: Reg. | O/T

Signature

Date

Daily Totals: Reg. / O/T

Remarks

Approved

Date

Total Hours

Direct Hours

Indirect: Admin/Leave

Source: Mock-up of a Time Record form from the forthcoming AIA book, **Standardized Accounting for Architects.** This publication and its related forms comprise the Manual Applications component of the AIA Financial Management System. This Time Record form, which can be used for both biweekly and semimonthly payrolls, is compatible with related forms of the AIA Computer-based Financial Management Service, thus easing the transition from manual to automated applications.

AIA FMS THE AMERICAN INSTITUTE OF ARCHITECTS FINANCIAL MANAGEMENT SYSTEM

and indirect costs. All accounting journals are also included to provide a firm base for eventual preparation of financial statements. Monitoring may also be done manually.

A third option is to adopt the *full system level* of CFMS. In that case the A/E's office receives a complete financial analysis for each accounting period, including such reports as Profit and Loss, Balance Sheet, Aged Accounts Receivable, Project Earnings, and Profit Planning. A variety of supplementary reports is available to provide additional office management information as desired.

THE TIME RECORD

A key element in reliably budgeting and monitoring compensation is the time record submitted at prescribed intervals by in-house staff. Time record forms are being developed for use with both the manual and automated versions of the AIA Financial Management System (Exhibit 6-2), and may be used as follows:

—In the first group of columns at the left side of the form, the individual enters the project code numbers;

—In the second group of columns, enter the code for the department or discipline (e.g. "architectural," "structural") in the first column; the project phase in the second column; the service codes to conform to the *Schedule of Designated Services* listing in the third and fourth columns; and the staff classification code in the fifth column.*

—Other information, such a project name and hours worked is entered in the appropriate line. Personnel are also identified, as are respective days and dates.

A/Es with a manual system are free to interchange the column arrangement, but at the risk of possibly costly readjustment should they go to the computer-based service.

Clearly, the *Guidelines* can serve the architect and engineer in a way that goes far beyond the basic role of determining compensation. Monitoring compensation is only one major way in which to put to use the important information developed under the *Guidelines*.

*See Exhibit A-1, Appendix A, for listing of suggested departmental categories and staff classifications.

7. Conclusion

A rational process has long been needed that will help the client and A/E identify, quickly and simply, the services required to carry out a project, assign responsibility for them to the client, A/E or others, and permit accurate estimates of the cost of providing these services. A fair and equitable approach to cost-based compensation designed to accommodate this broad set of needs has been described in these pages. It works equally well for small projects as for large ones and does these things:

—It contains a uniform listing (with descriptions) of virtually all the professional services that may be needed to carry out a project.

—It provides a clear understanding of the scope of services by setting up a simple process for the client and A/E to follow as they jointly identify the particular items of service required.

—It includes a way to assign responsibility for carrying out each such item.

—It offers a step-by-step procedure for the A/E to compute the amount of compensation for carrying out those services assigned to the A/E, based on personnel hours and expenses, plus an allowance for overhead and an amount for profit.

—It comes with a process the A/E may use to monitor costs once the program gets underway. This serves to reinforce businesslike administration of the client's project.

—It sets up a data base which may be used to reevaluate, if necessary, the services required as the project advances.

This cost-based compensation approach lets both the client and the A/E know exactly what services the estimated compensation will cover *before* any agreement is signed. With its consistent format and rational procedures, it is a solution aptly attuned to the tough, accountability-oriented demands of today's society.

Appendix A: Scope of Designated Services

The combinations of services required for specific projects are as varied as the projects themselves. The following list of services is offered to help determine and establish responsibilities for services required for a specific project.

Services are divided into nine phases. The first eight reflect sequential design and construction phases that conform to service groupings used in current documents. The ninth phase, entitled "Supplemental Services," covers services (such as expert witness or project public relations) for which need cannot be predetermined or scope determined, and which do not easily fall into one of the other eight phases or which span several phases.

Detailed descriptions have been developed for each service. These are included as part of AIA Document B162, Scope of Designated Services (see Appendix D).

A/Es should review the descriptions of items of service, those which suggest *potential* elements of service for each project, and adapt them as appropriate for their own firm and for specific proposals and cost estimates. *The lists and descriptions of services are not to be construed as standards of performance for the A/E on any particular project;* they are intended as tools for a better understanding between client and A/E as to the scope of services required, and to help the A/E in estimating the cost of those services.

Users of the *Guidelines* may supplement the printed list of services provided on the *Schedule of Designated Services* and *Phase Compensation Worksheet* forms by filling in blank spaces, and by deleting headings for services not required.

KEY TO THE NUMBER/LETTER CODE

To permit use with the Computer-based Financial Management Service (CFMS), a number/letter code has been developed to identify each discipline or department; phase and service; and staff level.

A typical entry might be:

A	4	28	3
1	2	3	4

The positions indicate, respectively—

Position #1: Department or discipline
(In this Example: A = Architectural)

Position #2: Phase
(In this Example: 4 = Design Development)

Position #3 (double digit): Item of service
(In this Example: 28 = Materials Research/Specifications)

Position #4: Staff/personnel designation
(In this Example: 3 = Technical I)

The line entry A4.28.3 therefore would signify time spent on materials research and development of the architectural outline specifications during design development by an upper level technical staff person. The digits are for use by the computer under CFMS and for general ease of identification. They may also be used on manual financial record systems.

For the appropriate department or discipline, and staff level classifications, see Exhibit A-1 for suggested categories.

SCHEDULE OF DESIGNATED SERVICES BY PHASE

The Phase/Service Matrix, AIA Form F860, Exhibit A-2, shows the interrelationship of the designated services by phases. By blacking the circle of selected services, the matrix may be used to identify those services on a specific project which will be estimated and on which personnel time will be recorded. Following is the list of designated services by phase:

Phase 1: Predesign Services

1.01 Project Administration
1.02 Disciplines Coordination/Document Checking
1.03 Agency Consulting/Review/Approval
1.04 Owner-supplied Data Coordination
1.05 Programming
1.06 Space Schematics/Flow Diagrams
1.07 Existing Facilities Surveys
1.08 Marketing Studies
1.09 Economic Feasibility Studies
1.10 Project Financing
1.29 Project Development Scheduling
1.31 Project Budgeting
1.33 Presentations

Phase 2: Site Analysis Services

2.01 Project Administration
2.02 Disciplines Coordination/Document Checking
2.03 Agency Consulting/Review/Approval
2.04 Owner-supplied Data Coordination
2.13 Site Analysis and Selection
2.14 Site Development Planning
2.15 Detailed Site Utilization Studies
2.16 On-site Utility Studies
2.17 Off-site Utility Studies
2.18 Environmental Studies and Reports
2.19 Zoning Processing Assistance
2.29 Project Development Scheduling
2.31 Project Budgeting
2.33 Presentations

Phase 3: Schematic Design Services

3.01 Project Administration

EXHIBIT A-1.

DEPARTMENT/DISCIPLINE DESIGNATIONS

A	Architectural	G	Graphics
S	Structural	P	Planning
M	Mechanical	U	Urban Design
E	Electrical	B	Estimating (Budgeting)
C	Civil	X	Specifications
L	Landscape Design	Y	Construction Administration
I	Interiors	Z	Special/Other

The above assumes that specialized services would be either assigned to a "regular" department (e.g., 'Lighting' to 'Electrical'; 'Acoustical' to 'Mechanical'; 'Kitchen' to 'Architectural' or 'Planning,' etc.) *or* indicated as a special department under "Z." Other codes may be added by individual offices to suit their needs. Where a department needs finer detail, it may be subdivided; example:

- M Mechanical: HVAC
- N Mechanical: Plumbing
- O Mechanical: Fire Protection

STAFF/PERSONNEL DESIGNATIONS

This is based upon the five categories indicated below. Examples of possible classifications of personnel for each category are indicated. Each firm must decide where its staff is assigned. Individuals may play different roles on different projects *as long as their billing rates* are accounted for properly.

1. PRINCIPAL (owner; partner; corporate officer; participating associate)
2. SUPERVISORY (general manager; department head; project manager; project architect; project engineer)
3. TECHNICAL I (senior professional staff—frequently licensed; highly skilled specialist; job captain; senior designer; senior drafter; senior planner; senior specifier; senior construction administrator)
4. TECHNICAL II (intermediate technical staff—usually not licensed—including intermediate levels of positions listed in Technical I; manager of clerical staff)
5. TECHNICAL III (junior technical staff—not licensed—including junior levels of positions listed in Technical I; secretarial or clerical staff; office assistant)

Staff categories are intended to establish groupings by salary levels rather than by function, primarily to assist in accurate budgeting and cost control monitoring.

3.02 Disciplines Coordination/Document Checking
3.03 Agency Consulting/Review/Approval
3.04 Owner-supplied Data Coordination
3.21 Architectural Design/Documentation
3.22 Structural Design/Documentation
3.23 Mechanical Design/Documentation
3.24 Electrical Design/Documentation
3.25 Civil Design/Documentation
3.26 Landscape Design/Documentation
3.27 Interior Design/Documentation
3.28 Materials Research/Specifications
3.29 Project Development Scheduling
3.32 Statement of Probable Construction Cost
3.33 Presentations

Phase 4: Design Development Services

4.01 Project Administration
4.02 Disciplines Coordination/Document Checking
4.03 Agency Consulting/Review/Approval
4.04 Owner-supplied Data Coordination
4.21 Architectural Design/Documentation
4.22 Structural Design/Documentation
4.23 Mechanical Design/Documentation
4.24 Electrical Design/Documentation
4.25 Civil Design/Documentation
4.26 Landscape Design/Documentation
4.27 Interior Design/Documentation
4.28 Materials Research/Specifications
4.29 Project Development Scheduling
4.32 Statement of Probable Construction Cost
4.33 Presentations

Phase 5: Construction Documents Services

5.01 Project Administration
5.02 Disciplines Coordination/Document Checking

EXHIBIT A-2. SAMPLE PARTIALLY COMPLETED PHASE/SERVICE MATRIX FORM F860

● = Time Estimates and Records ◐ = Time Estimates Only ○ = No Time Estimates or Records

(This example is completed for Phase 5 only, compatibly with Exhibit 4-3)

PHASE/SERVICE MATRIX

Project *Franklin Elementary* Project # *7701*

	PHASE 1: PREDESIGN SERVICES 1	PHASE 2: SITE ANALYSIS SERVICES 2	PHASE 3: SCHEMATIC DESIGN SERVICES 3	PHASE 4: DESIGN DEVELOPMENT SERVICES 4	PHASE 5: CONSTRUCTION DOCUMENTS SERVICES 5	PHASE 6: BIDDING OR NEGOTIATIONS SERVICES 6	PHASE 7: CONSTRUCTION CONTRACT ADMINISTRATION SERVICES 7	PHASE 8: POST-CONSTRUCTION SERVICES 8	PHASE 9: SUPPLEMENTAL SERVICES 9a	PHASE 9: (CONT'D) SUPPLEMENTAL SERVICES 9b	
A	○ .01 Project Administration	○ .01 Project Administration	○ .01 Project Administration	○ .01 Project Administration	● .01 Project Administration	○ .01 Project Administration	○ .01 Project Administration	○ .01 Project Administration	○ .61 Special Studies	○ .79 Materials and Systems Testing	A
B	○ .02 Disciplines Coordination/Document Checking	○ .02 Disciplines Coordination/Document Checking	○ .02 Disciplines Coordination/Document Checking	○ .02 Disciplines Coordination/Document Checking	● .02 Disciplines Coordination/Document Checking	○ .02 Disciplines Coordination/Document Checking	○ .02 Disciplines Coordination/Document Checking	○ .02 Disciplines Coordination/Document Checking	○ .62 Renderings	○ .80 Demolition Services	B
C	○ .03 Agency Consulting/Review/Approval	○ .03 Agency Consulting/Review/Approval	○ .03 Agency Consulting/Review/Approval	○ .03 Agency Consulting/Review/Approval	● .03 Agency Consulting/Review/Approval	○ .03 Agency Consulting/Review/Approval	○ .03 Agency Consulting/Review/Approval	○ .03 Agency Consulting/Review/Approval	○ .63 Model Construction	○ .81 Mock-up Services	C
D	○ .04 Owner-supplied Data Coordination	○ .04 Owner-supplied Data Coordination	○ .04 Owner-supplied Data Coordination	○ .04 Owner-supplied Data Coordination	● .04 Owner-supplied Data Coordination	○ .04 Owner-supplied Data Coordination	○ .04 Owner-supplied Data Coordination	○ .04 Owner-supplied Data Coordination	○ .64 Life Cycle Cost Analysis	○ .82 Still Photography	D
E	○ .05 Programming	○ .13 Site Analysis and Selection	○ .21 Architectural Design/Documentation	○ .21 Architectural Design/Documentation	● .21 Architectural Design/Documentation	○ .34 Bidding Materials	○ .41 Office Construction Administration	○ .50 Maintenance and Operational Programming	○ .65 Value Analysis	○ .83 Motion Pictures and Videotape	E
F	○ .06 Space Schematics/Flow Diagrams	○ .14 Site Development Planning	○ .22 Structural Design/Documentation	○ .22 Structural Design/Documentation	● .22 Structural Design/Documentation	○ .35 Addenda	○ .42 Construction Field Observation	○ .51 Start-up Assistance	○ .66 Quantity Surveys	○ .84 Coordination with Non-Design Professionals	F
G	○ .07 Existing Facilities Surveys	○ .15 Detailed Site Utilization Studies	○ .23 Mechanical Design/Documentation	○ .23 Mechanical Design/Documentation	◐ .23 Mechanical Design/Documentation	○ .36 Bidding/Negotiations	○ .43 Project Representation	○ .52 Record Drawings	○ .67 Detailed Construction Cost Estimates	○ .85 Special Disciplines Consultation	G
H	○ .08 Marketing Studies	○ .16 On-site Utility Studies	○ .24 Electrical Design/Documentation	○ .24 Electrical Design/Documentation	◐ .24 Electrical Design/Documentation	○ .37 Analysis of Alternates/Substitutions	○ .44 Inspection Coordination	○ .53 Warranty Review	○ .68 Energy Studies	○ .86 Special Building Type Consultation	H
I	○ .09 Economic Feasibility Studies	○ .17 Off-site Utility Studies	○ .25 Civil Design/Documentation	○ .25 Civil Design/Documentation	● .25 Civil Design/Documentation	○ .38 Special Bidding Services	○ .45 Supplemental Documents	○ .54 Postconstruction Evaluation	○ .69 Environmental Monitoring	○	I
J	○ .10 Project Financing	○ .18 Environmental Studies and Reports	○ .26 Landscape Design/Documentation	○ .26 Landscape Design/Documentation	● .26 Landscape Design/Documentation	○ .39 Bid Evaluation	○ .46 Quotation Requests/Change Orders	○	○ .70 Tenant-related Services	○	J
K	○	○ .19 Zoning Processing Assistance	○ .27 Interior Design/Documentation	○ .27 Interior Design/Documentation	● .27 Interior Design/Documentation	○ .40 Construction Contract Agreements	○ .47 Project Schedule Monitoring	○	○ .71 Graphics Design	○	K
L	○	○	○ .28 Materials Research/Specifications	○ .28 Materials Research/Specifications	● .28 Materials Research/Specifications	○	○ .48 Construction Cost Accounting	○	○ .72 Fine Arts and Crafts Services	○	L
M	○ .29 Project Development Scheduling	○ .29 Project Development Scheduling	○ .29 Project Development Scheduling	○ .29 Project Development Scheduling	○ .30 Special Bidding Documents/Scheduling	○	○ .49 Project Closeout	○	○ .73 Special Furnishings Design	○	M
N	○ .31 Project Budgeting	○ .31 Project Budgeting	○ .32 Statement of Probable Construction Cost	○ .32 Statement of Probable Construction Cost	● .32 Statement of Probable Construction Cost	○	○	○	○ .74 Non-Building Equipment Selection	○	N
O	○ .33 Presentations	○ .33 Presentations	○ .33 Presentations	○ .33 Presentations	● .33 Presentations	○	○	○	○ .75 Project Promotion/Public Relations	○	O
P	○	○	○	○	● *.71 Graphics Design*	○	○	○	○ .76 Leasing Brochures	○	P
Q	○	○	○	○	○	○	○	○	○ .77 Expert Witness	○	Q
R	○	○	○	○	○	○	○	○	○ .78 Computer Applications	○	R

AIA® FORM F860 • PHASE/SERVICE MATRIX • JANUARY 1978 EDITION • © 1978

THE AMERICAN INSTITUTE OF ARCHITECTS, 1735 NEW YORK AVE., N.W., WASHINGTON, DC 20006

THE AMERICAN INSTITUTE OF ARCHITECTS
FINANCIAL MANAGEMENT SYSTEM

5.03 Agency Consulting/Review/ Approval
5.04 Owner-supplied Data Coordination
5.21 Architectural Design/Documentation
5.22 Structural Design/Documentation
5.23 Mechanical Design/Documentation
5.24 Electrical Design/Documentation
5.25 Civil Design/Documentation
5.26 Landscape Design/Documentation
5.27 Interior Design/Documentation
5.28 Materials Research/Specifications
5.30 Special Bidding Documents/ Scheduling
5.32 Statement of Probable Construction Cost
5.33 Presentations

Phase 6: Bidding or Negotiations Services

6.01 Project Administration
6.02 Disciplines Coordination/ Document Checking
6.03 Agency Consulting/Review/ Approval
6.04 Owner-supplied Data Coordination
6.34 Bidding Materials
6.35 Addenda
6.36 Bidding/Negotiations
6.37 Analysis of Alternates/Substitutions
6.38 Special Bidding Services
6.39 Bid Evaluation
6.40 Construction Contract Agreements

Phase 7: Construction Contract Administration Services

7.01 Project Administration
7.02 Disciplines Coordination/Document Checking
7.03 Agency Consulting/Review/ Approval
7.04 Owner-supplied Data Coordination
7.41 Office Construction Administration
7.42 Construction Field Observation
7.43 Project Representation
7.44 Inspection Coordination
7.45 Supplemental Documents
7.46 Quotation Requests/Change Orders
7.47 Project Schedule Monitoring
7.48 Construction Cost Accounting
7.49 Project Closeout

Phase 8: Postconstruction Services

8.01 Project Administration
8.02 Disciplines Coordination/ Document Checking
8.03 Agency Consulting/Review/ Approval
8.04 Owner-supplied Data Coordination
8.50 Maintenance and Operational Programming
8.51 Start-up Assistance
8.52 Record Drawings
8.53 Warranty Review
8.54 Postconstruction Evaluation

Phase 9: Supplemental Services

9.61 Special Studies
9.62 Renderings
9.63 Model Construction
9.64 Life Cycle Cost Analysis
9.65 Value Analysis
9.66 Quantity Surveys
9.67 Detailed Construction Cost Estimates
9.68 Energy Studies
9.69 Environmental Monitoring
9.70 Tenant-related Services
9.71 Graphics Design
9.72 Fine Arts and Crafts Services
9.73 Special Furnishings Design
9.74 Non-Building Equipment Selection
9.75 Project Promotion/Public Relations
9.76 Leasing Brochures
9.77 Expert Witness
9.78 Computer Applications
9.79 Materials and Systems Testing
9.80 Demolition Services
9.81 Mock-up Services
9.82 Still Photography
9.83 Motion Pictures and Videotape
9.84 Coordination with Non-Design Professionals
9.85 Special Disciplines Consultation
9.86 Special Building Type Consultation

DETAILED DESCRIPTIONS OF DESIGNATED SERVICES

Detailed descriptions of designated services are included as part of AIA Document B162. See Appendix D.

Appendix B: Methods of Compensation

This appendix reviews several methods of compensation in current use. Whatever method or combination of methods the client and A/E agree upon, it is essential for the A/E to have developed, as a result of following the procedures in the *Guidelines,* compensation amounts for total estimated costs and planned profit as well as an amount for estimated reimbursables. For many clients, especially federal government agencies, the A/E must present an indirect expense factor and be ready to furnish appropriate backup information.

CRITERIA FOR EVALUATING METHODS

The method or methods chosen depend on a number of factors. The following criteria are commonly used in evaluating compensation methods:

—Does it permit the A/E to recover costs and net an adequate profit, assuming reasonably efficient office and project management?

—Does it accurately reflect to the client the value of the A/E's services?

—Does it provide for compensation increases brought about because of changes in the scope of services or factors outside the A/E's control, such as strikes, client delays?

—Does it provide for compensation increases due to rising costs of providing services during the course of a long project, such as staff salary increases?

—Could it create legal problems?

—Does it encourage the client to cooperate fully in advancing the project?

—Does it act as an incentive to draw from the A/E high-quality design, technical competence and managerial efficiency?

—Is it easy to understand and simple to use?

—Does it allow the client to predetermine satisfactorily the costs for the A/E services?

The A/E may think of other criteria. No matter what the method(s) chosen, no one method will meet all the criteria; some criteria may be more applicable to a specific commission than others. For example, a complex renovation project, with many unknowns, may best be handled on an hourly or multiple of direct personnel expense basis rather than using a percentage of construction cost approach.

Accordingly, the A/E and client must attach weights to the criteria. These will depend on such factors as the length of the client-A/E relationship and degree of trust; the nature of the work the A/E firm generally performs; the type of project planned; the duration of construction; and the likelihood and extent of changes in the scope of work. This "weighting" is discussed in more detail in the latter part of this chapter.

REVIEW OF METHODS OF COMPENSATION

Following is a review of the most common methods of compensation (see also Exhibit B-1):

A. Multiple of Direct Salary Expense

This method of cost-based compensation is related directly to the amount of time and effort required to provide professional services for a project. It is one of several methods well suited to large, complex projects, those in which the scope is undefined or not well defined, and those requiring special or supplemental services. Under this method, direct salary expense incurred on the project is multiplied by a factor. This factor is not the same as the indirect expense factor used in the *Guidelines,* since it must account not only for indirect expense but also other nonreimbursable direct expense and profit. A variant involves payment for principals' time at a fixed hourly rate and employees' time at a multiple of direct salary or direct personnel expense. Reimbursable expense is accounted for separately, as is outside services expense. In choosing the multiple, it is therefore important to analyze *all* elements contributing to this factor to be sure that all items of expense and profit are included (see also Chapter 10 of Reference 5, Appendix C for a broader discussion of the multiple).

B. Multiple of Direct Personnel Expense

This method of cost-based compensation is similar to multiple of direct salary expense. As direct personnel expense comprises both direct salaries and direct benefits, the multiple used for this method would be lower than for method A, reflecting a lower indirect expense factor. However, the amounts billed to the client under either method A or B should be approximately the same for the hours worked. Refer to the discussion in Chapters 2 and 4.

C. Professional Fee Plus Expenses

This is another cost-based method of compensation, similar to Multiple of Direct Salary or Direct Personnel Expense. Under this method, the multiple does not include a factor for profit and is therefore lower, reflecting only expense. Alternatively, all items of incurred direct expense and pro rata share of indirect expense may be made reimbursable. Planned profit (professional fee) may be a stipulated sum, or it may be a percentage of total expense or of construction cost; it is often set in advance of services performed.

D. Percentage of Construction Cost

This method is not based directly on the A/E's costs but on an assumed definitive correlation between the construction cost of a project and the A/E's efforts and expertise. Compensation is determined by applying a percentage scale against

EXHIBIT B-1. COMPENSATION METHODS

Fixed	Cost Plus	Unit	Repetitive
D. % of Construction Cost	A.B. Multiple of Direct Salary or DPE C. Professional Fee (Planned Profit) + Expenses	H1. Square Foot	H3. Royalty
E. Stipulated Sum	F. Hourly (or per diem) G. Multiple of Amounts Billed to Architect	H2. Unit Cost	Various methods for repeat work
H4. Combined Methods			

actual construction cost; the actual percentage used is a variable dependent on the nature of the project (its type, size, cost, quality, complexity), the scope of services to be rendered, and the method of awarding the construction contract or contracts. An inherent inequity of this method for both client and A/E is its lack of incentive to the A/E to reduce the project's construction cost. In other words, the method may bear little relationship to the actual effort expended or the expertise of the A/E firm.

If this percentage method must be used, the fairest approach is to establish the percentage figure only after the work is at a stage at which the construction cost can be estimated with reasonable accuracy; prior to this point, payments on account of compensation are best calculated on a "cost-plus" basis.

When the method is used, it is a good plan to complete the Scope of Designated Services process (see Chapter 3), analyze the work sheet calculations (Chapter 4), and convert the result to a percentage, using the backup work as support documentation for negotiation.

E. Stipulated Sum

If the scope and schedule of the project are sufficiently defined at the outset to allow the A/E to calculate proposed compensation accurately, the client and A/E may agree on a stipulated sum for compensation. To avoid misunderstanding, services to be provided should be clearly listed and described in the agreement. An alternative is to use a combined method: traditional services provided under a stipulated sum, all other services by "cost-plus."

F. Hourly Billing Rates

This cost-based method of compensation, not unlike the Multiple of Direct Salary or Direct Personnel Expense methods, is suitable where the time required to solve a problem or set of problems is hard to assess. Hourly billing rates are established for each level of in-house personnel, or for specific personnel to account for direct personnel expense, pro rata share of indirect expense, other direct nonreimbursable expense, and planned profit. Outside services expense may be billed by stipulated sum as expense is incurred or, like in-house expenses, may be accounted for by levels of personnel or specific personnel of outside services contractors. Billings for outside services expense should be marked up, as should other reimbursable expense, to account for the A/E's coordination (except where it is included in in-house billings), handling expenses and risk.

A variant is to use daily billing rates in lieu of hourly. Long-term agreements should contain a provision for periodic adjustment of hourly or daily rates to reflect changes in the A/E's costs.

G. Multiple of Amounts Billed to Architect

This method of compensation is suitable to situations where the architect engages and coordinates the work of outside services contractors or professional consultants. See discussion under Hourly Billing Rates concerning markup.

H. Other

1. Square footage. Compensation is computed by multiplying the square footage (gross area) of a building by a price factor. This method lends itself to large, simple projects (such as warehouses) for which the scope and complexity can be clearly defined. Not in general use by many firms, it depends on experience on similar work for accurate judgment of the services and compensation required. A variant on this method is compensation per net square foot of a building, sometimes used for layout of tenant spaces in shopping centers and office buildings.

2. Unit cost. Compensation based on unit costs may be calculated for a number of project types. Services required are usually in direct proportion to the size of the project and may be compensated as follows:

—For motels/hotels: cost per room
—For apartments: cost per unit (may vary with unit type or mix)
—For land planning: cost per acre

As with the square footage method, experience is required for accurate estimates. One hazard lies in failing to recognize differences in complexity of nonrepetitive portions of the project. For example, a commercial roadside motel may have only 5% or 10% of its room area devoted to public space, whereas for a convention or resort hotel, this figure may run from 40% to 200%. Similarly, apartment buildings may have sharply varying sizes of units and amounts of space devoted to common use. This method is best used in combination with other methods.

3. Royalty. Royalty arrangements apply in such cases as repeats of a basic building type, as discussed further in the later section on repeat work. Agreements must be tailor-made to the project. In arrangements of this type, the A/E has two options (of which there may be variants):

i) To set compensation for the initial work based on the expectation of enough later royalty income to produce a profit on the entire project. In this case the degree of risk must be measured and the royalties set accordingly.
ii) To recover costs and planned profit on the prototype and receive royalties for reuse of the design as well as compensation for extra work and related profit. This is a more sensible business arrangement.

Another royalty situation may occur when the A/E is to provide repeat services for a building designed by another firm. Thus, a franchiser who has an A/E-designed prototype building may expand into states where this A/E is not registered or the site is too remote for the A/E to service the project efficiently. This may require the services of another practitioner. In this type of arrangement, care should be taken to define responsibilities, especially as to risk of liability. A valid hold-harmless agreement is applied only to contracting parties and does not preclude a third-party suit. Its value depends on the indemnifying party's financial responsibility. Consult with legal counsel and the insurer prior to commitment.

4. Combined methods. On large, complex, ill-defined or certain special projects, a combination of methods becomes viable. The A/E may tailor a proposal to the specific problem. For example, a large low-rise, highly repetitive multifamily project, including land planning, might be approached as follows:

—Land planning phase: cost per acre
—Design of typical apartment units: hourly or "cost-plus" type method
—Construction drawings for typical units and buildings: stipulated sum
—Repeats and site adaptation: unit cost

Estimating costs in this manner is quite simple and will usually entail fewer problems than trying to estimate phases by inappropriate methods. By combining methods, the A/E may estimate compensation for each phase by the most sensible method. However, care should be taken to avoid a line-by-line "shopping list" of diverse services and compensation methods.

Many firms contract for certain early phases on an hourly arrangement and for the balance by percentage of construction cost, stipulated sum, or other method, with amounts determined as the scope becomes defined. Since most unknowns occur during the early stages of the A/E's work, this approach is fair to both parties. The A/E is assured of time in the design phases to solve the problem; the client need not contract for all of the A/E's services on the more open-ended, "cost-plus" basis.

REPEAT WORK

There are several applications of methods of compensation for projects involving repeat work. Several examples are:

—*The client wishes to build a second building the same as or similar to the original, on a different site.* In this instance, compensation for subsequent buildings should be negotiated so as to fairly reflect expended time, profit and expense for any changes to and reuse of original contract documents and adaptation to the new site. Most applicable are "cost-plus" types or stipulated sum.

—*The client wishes to build a large multi-family development.* Here, a combination of methods best recognizes the differing complexities of each phase. The work required in schematic design and design development may depend on the number of typical units required. For example, work required in the design phase for 200 three-bedroom apartments of the same size and configuration may be far less than that for 200 units with many suite sizes and types. A combined method would use the unit cost method for the design of each type of unit, and another method for other phases. The same applies to motels, office and industrial parks.

—*The client wishes to build a large, repetitive development in stages over several years.* Compensation for the first stage should cover all basic design and construction documents costs, with later stages compensated by a "cost-plus" or stipulated sum method. Occasionally, however, the client must negotiate on the basis of equal dollar amounts becoming available at each stage, despite the disproportionate cost required for the *initial* stage. The A/E should establish a "bail-out" cost to at least cover first stage expenses should the project be dropped after the first stage. In any event, higher profits are justified to reflect the A/E's higher risk and carrying of receivables.

—*When a client embarks on a large scale program, such as a franchise operation, the royalty method is appropriate.* The client's enthusiasm at this point does not guarantee success, and the A/E is best advised to create a no-loss situation for the prototype in case the client decides to drop the construction program. The rewards for the risk undertaken should be proportionately high.

Compensation to account for recovery of these costs will be in addition to any royalties. The royalty should reflect compensation for reuse of the design and profit to the A/E. Alternatively, a stipulated sum or unit cost per building may be negotiated. The cost of administering contracts in a franchise-type situation is difficult to estimate accurately, especially if the geographic area is large.

Appendix C: Related References

1. "Agree Cautiously," by Thomas Nathan. *The AIA Journal,* November 1973.

2. "Allowable Overhead Costs," by McLane Fisher. *Professional Services Management Journal (PSMJ),* January 1976. (*PSMJ*, P.O. Box 11316, Newington, CT 06111).

3. *AIA Computer-based Financial Management Service.* For descriptive information write to: Dr. G. Neil Harper, Harper and Shuman, Inc., 1278 Massachusetts Ave., Cambridge, MA 02138.

4. *Contracting With The Federal Government* (A Primer for Architects/Engineers). COFPAES, 9170 Brookville Road, Silver Spring, MD 20910, 1974. $10.00 prepaid.

5. *Current Techniques in Architectural Practice,* edited by Robert Allan Class and Robert E. Koehler. Washington, DC: The American Institute of Architects, and New York: Architectural Record, 1976. $25.00 (AIA Catalog Number 2M701).

6. *The Federal Marketplace: Are You Prepared?* Washington, DC: The American Institute of Architects, 1976. $9.00 (AIA Catalog Number 2M704).

7. "Fee Negotiation with the Feds," by Michael Hough. *PSMJ*, January 1976.

8. "Maximizing Profit on Federal Contracts," by Michael Hough. *PSMJ*, March 1976.

9. *Standardized Accounting for Architects.* Washington, DC: The American Institute of Architects, scheduled for publication in mid-1978.

Appendix D: Forms

One complete sample set of the AIA forms recommended for use with the *Guidelines* is included, as enumerated below. Earlier chapters provide detailed directions and commentary as to the use of the forms. Additional copies of the forms are available from the same sales sources that carry this publication. Contact the engineering societies for their forms.

AIA Form or Document No.	Title
F810	Schedule of Designated Services Worksheet (blank)
F811	Schedule of Designated Services Worksheet, Phase 1
F812	Schedule of Designated Services Worksheet, Phase 2
F813	Schedule of Designated Services Worksheet, Phase 3
F814	Schedule of Designated Services Worksheet, Phase 4
F815	Schedule of Designated Services Worksheet, Phase 5
F816	Schedule of Designated Services Worksheet, Phase 6
F817	Schedule of Designated Services Worksheet, Phase 7
F818	Schedule of Designated Services Worksheet, Phase 8
F819	Schedule of Designated Services Worksheet, Phase 9
F820	Phase Compensation Worksheet (blank)
F821	Phase Compensation Worksheet, Phase 1
F822	Phase Compensation Worksheet, Phase 2
F823	Phase Compensation Worksheet, Phase 3
F824	Phase Compensation Worksheet, Phase 4
F825	Phase Compensation Worksheet, Phase 5
F826	Phase Compensation Worksheet, Phase 6
F827	Phase Compensation Worksheet, Phase 7
F828	Phase Compensation Worksheet, Phase 8
F829	Phase Compensation Worksheet, Phase 9
F830	Project Compensation Worksheet
F840	Project Compensation Summary
F850	Project Time and Payment Schedule
F860	Phase/Service Matrix
B161	Standard Form of Agreement Between Owner and Architect for Designated Services
B161a	Instruction Sheet for AIA Document B161
B162	Scope of Designated Services (for use with B161)
B162a	Instruction Sheet for AIA Document B162

SCHEDULE OF DESIGNATED SERVICES WORKSHEET

Project ______________________________

Project # ________________ Date ____________

Owner ______________________________

Architect ______________________________

	By Architect	By Architect, as Outside Services			By Owner, Coordinated by Architect	By Owner			By Architect, as Additional Service*	Not to be Provided	Method of Compensation

METHODS OF COMPENSATION

A = Multiple of Direct Salary Expense
B = Multiple of Direct Personnel Expense
C = Professional Fee Plus Expenses
D = Percentage of Construction Cost
E = Stipulated Sum
F = Hourly Billing Rates
G = Multiple of Amounts Billed to Architect
H = Other:

*Requires separate authorization and compensation

SCHEDULE OF DESIGNATED SERVICES WORKSHEET **PHASE 1: PREDESIGN SERVICES**

Project ____________________

Project # __________ Date __________

Owner ____________________

Architect ____________________

	By Architect	By Architect, as Outside Services			By Owner, Coordinated by Architect	By Owner			By Architect, as Additional Service*	Not to be Provided	Method of Compensation
1.01 Project Administration											
1.02 Disciplines Coordination/Document Checking											
1.03 Agency Consulting/Review/Approval											
1.04 Owner-supplied Data Coordination											
1.05 Programming											
1.06 Space Schematics/Flow Diagrams											
1.07 Existing Facilities Surveys											
1.08 Marketing Studies											
1.09 Economic Feasibility Studies											
1.10 Project Financing											
1.29 Project Development Scheduling											
1.31 Project Budgeting											
1.33 Presentations											

METHODS OF COMPENSATION

A = Multiple of Direct Salary Expense
B = Multiple of Direct Personnel Expense
C = Professional Fee Plus Expenses
D = Percentage of Construction Cost
E = Stipulated Sum
F = Hourly Billing Rates
G = Multiple of Amounts Billed to Architect
H = Other:

*Requires separate authorization and compensation

SCHEDULE OF DESIGNATED SERVICES WORKSHEET **PHASE 2: SITE ANALYSIS SERVICES**

Project ______________________

Project # ______________ Date ______________

Owner ______________________

Architect ______________________

	By Architect	By Architect, as Outside Services			By Owner, Coordinated by Architect	By Owner			By Architect, as Additional Service*	Not to be Provided	Method of Compensation
2.01 Project Administration											
2.02 Disciplines Coordination/Document Checking											
2.03 Agency Consulting/Review/Approval											
2.04 Owner-supplied Data Coordination											
2.13 Site Analysis and Selection											
2.14 Site Development Planning											
2.15 Detailed Site Utilization Studies											
2.16 On-site Utility Studies											
2.17 Off-site Utility Studies											
2.18 Environmental Studies and Reports											
2.19 Zoning Processing Assistance											
2.29 Project Development Scheduling											
2.31 Project Budgeting											
2.33 Presentations											

METHODS OF COMPENSATION

A = Multiple of Direct Salary Expense
B = Multiple of Direct Personnel Expense
C = Professional Fee Plus Expenses
D = Percentage of Construction Cost
E = Stipulated Sum
F = Hourly Billing Rates
G = Multiple of Amounts Billed to Architect
H = Other:

*Requires separate authorization and compensation

SCHEDULE OF DESIGNATED SERVICES WORKSHEET — **PHASE 3: SCHEMATIC DESIGN SERVICES**

Project ______________________

Project # ______________ Date ______________

Owner ______________________

Architect ______________________

	By Architect	By Architect, as Outside Services			By Owner, Coordinated by Architect	By Owner			By Architect, as Additional Service*	Not to be Provided	Method of Compensation
3.01 Project Administration											
3.02 Disciplines Coordination/Document Checking											
3.03 Agency Consulting/Review/Approval											
3.04 Owner-supplied Data Coordination											
3.21 Architectural Design/Documentation											
3.22 Structural Design/Documentation											
3.23 Mechanical Design/Documentation											
3.24 Electrical Design/Documentation											
3.25 Civil Design/Documentation											
3.26 Landscape Design/Documentation											
3.27 Interior Design/Documentation											
3.28 Materials Research/Specifications											
3.29 Project Development Scheduling											
3.32 Statement of Probable Construction Cost											
3.33 Presentations											

METHODS OF COMPENSATION

A = Multiple of Direct Salary Expense
B = Multiple of Direct Personnel Expense
C = Professional Fee Plus Expenses
D = Percentage of Construction Cost
E = Stipulated Sum
F = Hourly Billing Rates
G = Multiple of Amounts Billed to Architect
H = Other:

*Requires separate authorization and compensation

SCHEDULE OF DESIGNATED SERVICES WORKSHEET

PHASE 4: DESIGN DEVELOPMENT SERVICES

Project ________ Project # ________ Date ________ Owner ________ Architect ________	By Architect	By Architect, as Outside Services			By Owner, Coordinated by Architect	By Owner			By Architect, as Additional Service*	Not to be Provided	Method of Compensation
4.01 Project Administration											
4.02 Disciplines Coordination/Document Checking											
4.03 Agency Consulting/Review/Approval											
4.04 Owner-supplied Data Coordination											
4.21 Architectural Design/Documentation											
4.22 Structural Design/Documentation											
4.23 Mechanical Design/Documentation											
4.24 Electrical Design/Documentation											
4.25 Civil Design/Documentation											
4.26 Landscape Design/Documentation											
4.27 Interior Design/Documentation											
4.28 Materials Research/Specifications											
4.29 Project Development Scheduling											
4.32 Statement of Probable Construction Cost											
4.33 Presentations											

METHODS OF COMPENSATION

A = Multiple of Direct Salary Expense
B = Multiple of Direct Personnel Expense
C = Professional Fee Plus Expenses
D = Percentage of Construction Cost
E = Stipulated Sum
F = Hourly Billing Rates
G = Multiple of Amounts Billed to Architect
H = Other:

*Requires separate authorization and compensation

SCHEDULE OF DESIGNATED SERVICES WORKSHEET **PHASE 5: CONSTRUCTION DOCUMENTS SERVICES**

Project ______ Project # ______ Date ______ Owner ______ Architect ______	By Architect	By Architect, as Outside Services			By Owner, Coordinated by Architect	By Owner			By Architect, as Additional Service*	Not to be Provided	Method of Compensation
5.01 Project Administration											
5.02 Disciplines Coordination/Document Checking											
5.03 Agency Consulting/Review/Approval											
5.04 Owner-supplied Data Coordination											
5.21 Architectural Design/Documentation											
5.22 Structural Design/Documentation											
5.23 Mechanical Design/Documentation											
5.24 Electrical Design/Documentation											
5.25 Civil Design/Documentation											
5.26 Landscape Design/Documentation											
5.27 Interior Design/Documentation											
5.28 Materials Research/Specifications											
5.30 Special Bidding Documents/Scheduling											
5.32 Statement of Probable Construction Cost											
5.33 Presentations											

METHODS OF COMPENSATION

A = Multiple of Direct Salary Expense
B = Multiple of Direct Personnel Expense
C = Professional Fee Plus Expenses
D = Percentage of Construction Cost
E = Stipulated Sum
F = Hourly Billing Rates
G = Multiple of Amounts Billed to Architect
H = Other:

*Requires separate authorization and compensation

SCHEDULE OF DESIGNATED SERVICES WORKSHEET	PHASE 6: BIDDING OR NEGOTIATIONS SERVICES

Project ______

Project # ______ Date ______

Owner ______

Architect ______

	By Architect	By Architect, as Outside Services			By Owner, Coordinated by Architect	By Owner			By Architect, as Additional Service*	Not to be Provided	Method of Compensation
6.01 Project Administration											
6.02 Disciplines Coordination/Document Checking											
6.03 Agency Consulting/Review/Approval											
6.04 Owner-supplied Data Coordination											
6.34 Bidding Materials											
6.35 Addenda											
6.36 Bidding/Negotiations											
6.37 Analysis of Alternates/Substitutions											
6.38 Special Bidding Services											
6.39 Bid Evaluation											
6.40 Construction Contract Agreements											

METHODS OF COMPENSATION

A = Multiple of Direct Salary Expense
B = Multiple of Direct Personnel Expense
C = Professional Fee Plus Expenses
D = Percentage of Construction Cost
E = Stipulated Sum
F = Hourly Billing Rates
G = Multiple of Amounts Billed to Architect
H = Other:

*Requires separate authorization and compensation

THE AMERICAN INSTITUTE OF ARCHITECTS
FINANCIAL MANAGEMENT SYSTEM

SCHEDULE OF DESIGNATED SERVICES WORKSHEET

PHASE 7: CONSTRUCTION CONTRACT ADMINISTRATION SERVICES

Project ______ Project # ______ Date ______ Owner ______ Architect ______	By Architect	By Architect, as Outside Services			By Owner, Coordinated by Architect	By Owner			By Architect, as Additional Service*	Not to be Provided	Method of Compensation
7.01 Project Administration											
7.02 Disciplines Coordination/Document Checking											
7.03 Agency Consulting/Review/Approval											
7.04 Owner-supplied Data Coordination											
7.41 Office Construction Administration											
7.42 Construction Field Observation											
7.43 Project Representation											
7.44 Inspection Coordination											
7.45 Supplemental Documents											
7.46 Quotation Requests/Change Orders											
7.47 Project Schedule Monitoring											
7.48 Construction Cost Accounting											
7.49 Project Closeout											

METHODS OF COMPENSATION

A = Multiple of Direct Salary Expense
B = Multiple of Direct Personnel Expense
C = Professional Fee Plus Expenses
D = Percentage of Construction Cost
E = Stipulated Sum
F = Hourly Billing Rates
G = Multiple of Amounts Billed to Architect
H = Other:

*Requires separate authorization and compensation

SCHEDULE OF DESIGNATED SERVICES WORKSHEET **PHASE 8: POSTCONSTRUCTION SERVICES**

Project ____ Project # ____ Date ____ Owner ____ Architect ____	By Architect	By Architect, as Outside Services			By Owner, Coordinated by Architect	By Owner			By Architect, as Additional Service*	Not to be Provided	Method of Compensation
8.01 Project Administration											
8.02 Disciplines Coordination/Document Checking											
8.03 Agency Consulting/Review/Approval											
8.04 Owner-supplied Data Coordination											
8.50 Maintenance and Operational Programming											
8.51 Start-up Assistance											
8.52 Record Drawings											
8.53 Warranty Review											
8.54 Postconstruction Evaluation											

METHODS OF COMPENSATION

A = Multiple of Direct Salary Expense
B = Multiple of Direct Personnel Expense
C = Professional Fee Plus Expenses
D = Percentage of Construction Cost
E = Stipulated Sum
F = Hourly Billing Rates
G = Multiple of Amounts Billed to Architect
H = Other:

*Requires separate authorization and compensation

SCHEDULE OF DESIGNATED SERVICES WORKSHEET

PHASE 9: SUPPLEMENTAL SERVICES (first of two pages)

Project ______________________________

Project # ______________ Date ______________

Owner ______________________________

Architect ______________________________

	By Architect	By Architect, as Outside Services			By Owner, Coordinated by Architect	By Owner			By Architect, as Additional Service*	Not to be Provided	Method of Compensation
9.61 Special Studies											
9.62 Renderings											
9.63 Model Construction											
9.64 Life Cycle Cost Analysis											
9.65 Value Analysis											
9.66 Quantity Surveys											
9.67 Detailed Construction Cost Estimates											
9.68 Energy Studies											
9.69 Environmental Monitoring											
9.70 Tenant-related Services											
9.71 Graphics Design											
9.72 Fine Arts and Crafts Services											
9.73 Special Furnishings Design											
9.74 Non-Building Equipment Selection											
9.75 Project Promotion/Public Relations											
9.76 Leasing Brochures											
9.77 Expert Witness											
9.78 Computer Applications											

METHODS OF COMPENSATION

A = Multiple of Direct Salary Expense
B = Multiple of Direct Personnel Expense
C = Professional Fee Plus Expenses
D = Percentage of Construction Cost
E = Stipulated Sum
F = Hourly Billing Rates
G = Multiple of Amounts Billed to Architect
H = Other:

*Requires separate authorization and compensation

SCHEDULE OF DESIGNATED SERVICES WORKSHEET — PHASE 9: SUPPLEMENTAL SERVICES (second of two pages)

Project ______________________

Project # ______________ Date ______________

Owner ______________________

Architect ______________________

	By Architect	By Architect, as Outside Services			By Owner, Coordinated by Architect	By Owner			By Architect, as Additional Service*	Not to be Provided	Method of Compensation
9.79 Materials and Systems Testing											
9.80 Demolition Services											
9.81 Mock-up Services											
9.82 Still Photography											
9.83 Motion Pictures and Videotape											
9.84 Coordination with Non-Design Professionals											
9.85 Special Disciplines Consultation											
9.86 Special Building Type Consultation											

METHODS OF COMPENSATION

A = Multiple of Direct Salary Expense
B = Multiple of Direct Personnel Expense
C = Professional Fee Plus Expenses
D = Percentage of Construction Cost
E = Stipulated Sum
F = Hourly Billing Rates
G = Multiple of Amounts Billed to Architect
H = Other:

*Requires separate authorization and compensation

PHASE COMPENSATION WORKSHEET

SERVICE																
IN-HOUSE PERSONNEL Hrs.																
$																
Hrs.																
$																
Hrs.																
$																
Hrs.																
$																
Hrs.																
$																
SUB TOT. Hrs.																
$																
OUT-SIDE Hrs.																
$																
TOT. Hrs.																
$																

Project________________________

Project #______________ Date__________

Owner________________________

Architect________________________

TOTAL HOURS	TOTAL DOLLARS	ITEM	LINE
		@ $	1
		@ $	2
		@ $	3
		@ $	4
		@ $	5
		Direct In-house Salary Expense	6
		Direct Personnel Expense	7
		Indirect Expense	8
		Other Nonreimbursable Direct Expense	9
		Total In-house Expense	10
		Outside Services Expense	11
		Estimated Total Expense	12
		Contingency	13
		Profit	14
		Proposed Compensation	15
		Estimated Reimbursable Expense	16

REMARKS

PHASE COMPENSATION WORKSHEET

SERVICE		1.01 Project Administration	1.02 Disciplines Coord./ Document Checking	1.03 Agency Consulting/ Review/Approval	1.04 Owner-supplied Data Coordination	1.05 Programming	1.06 Space Schematics/ Flow Diagrams	1.07 Existing Facilities Surveys	1.08 Marketing Studies	1.09 Economic Feasibility Studies	1.10 Project Financing	1.29 Project Development Scheduling	1.31 Project Budgeting	1.33 Presentations			
IN-HOUSE PERSONNEL	Hrs.																
	$																
	Hrs.																
	$																
	Hrs.																
	$																
	Hrs.																
	$																
	Hrs.																
	$																
SUB TOT.	Hrs.																
	$																
OUT-SIDE	Hrs.																
	$																
TOT.	Hrs.																
	$																

REMARKS

PHASE 1: PREDESIGN SERVICES

Project____________________

Project #____________ Date__________

Owner____________________

Architect____________________

TOTAL HOURS	TOTAL DOLLARS	ITEM	LINE
		@ $	1
		@ $	2
		@ $	3
		@ $	4
		@ $	5
		Direct In-house Salary Expense	6
		Direct Personnel Expense	7
		Indirect Expense	8
		Other Nonreimbursable Direct Expense	9
		Total In-house Expense	10
		Outside Services Expense	11
		Estimated Total Expense	12
		Contingency	13
		Profit	14
		Proposed Compensation	15
		Estimated Reimbursable Expense	16

THE AMERICAN INSTITUTE OF ARCHITECTS
FINANCIAL MANAGEMENT SYSTEM

PHASE COMPENSATION WORKSHEET

PHASE 2: SITE ANALYSIS SERVICES

Project ____________________

Project # ____________ Date ____________

Owner ____________________

Architect ____________________

SERVICE		2.01	2.02	2.03	2.04	2.13	2.14	2.15	2.16	2.17	2.18	2.19	2.29	2.31	2.33		
		Project Administration	Disciplines Coord./ Document Checking	Agency Consulting/ Review/Approval	Owner-supplied Data Coordination	Site Analysis and Selection	Site Development Planning	Detailed Site Utilization Studies	On-site Utility Studies	Off-site Utility Studies	Environmental Studies and Reports	Zoning Processing Assistance	Project Development Scheduling	Project Budgeting	Presentations		
IN-HOUSE PERSONNEL	Hrs.																
	$																
	Hrs.																
	$																
	Hrs.																
	$																
	Hrs.																
	$																
	Hrs.																
	$																
SUB TOT.	Hrs.																
	$																
OUT-SIDE	Hrs.																
	$																
TOT.	Hrs.																
	$																

TOTAL HOURS	TOTAL DOLLARS	ITEM	LINE
		@ $	1
		@ $	2
		@ $	3
		@ $	4
		@ $	5
		Direct In-house Salary Expense	6
		Direct Personnel Expense	7
		Indirect Expense	8
		Other Nonreimbursable Direct Expense	9
		Total In-house Expense	10
		Outside Services Expense	11
		Estimated Total Expense	12
		Contingency	13
		Profit	14
		Proposed Compensation	15
		Estimated Reimbursable Expense	16

REMARKS

PHASE COMPENSATION WORKSHEET

PHASE 3: SCHEMATIC DESIGN SERVICES

Project ______________________

Project # ______________ Date ______________

Owner ______________________

Architect ______________________

SERVICE		3.01 Project Administration	3.02 Disciplines Coord./ Document Checking	3.03 Agency Consulting/ Review/Approval	3.04 Owner-supplied Data Coordination	3.21 Architectural Design/ Documentation	3.22 Structural Design/ Documentation	3.23 Mechanical Design/ Documentation	3.24 Electrical Design/ Documentation	3.25 Civil Design/ Documentation	3.26 Landscape Design/ Documentation	3.27 Interior Design/ Documentation	3.28 Materials Research/ Specifications	3.29 Project Development Scheduling	3.32 Statement of Probable Construction Cost	3.33 Presentations		TOTAL HOURS	TOTAL DOLLARS	ITEM	LINE
IN-HOUSE PERSONNEL	Hrs.																			@ $	1
	$																				
	Hrs.																			@ $	2
	$																				
	Hrs.																			@ $	3
	$																				
	Hrs.																			@ $	4
	$																				
	Hrs.																			@ $	5
	$																				
SUB TOT.	Hrs.																			Direct In-house Salary Expense	6
	$																				
																				Direct Personnel Expense	7
																				Indirect Expense	8
																				Other Nonreimbursable Direct Expense	9
																				Total In-house Expense	10
OUT-SIDE	Hrs.																			Outside Services Expense	11
	$																				
TOT.	Hrs.																			Estimated Total Expense	12
	$																				

REMARKS	TOTAL DOLLARS	ITEM	LINE
		Contingency	13
		Profit	14
		Proposed Compensation	15
		Estimated Reimbursable Expense	16

THE AMERICAN INSTITUTE OF ARCHITECTS, 1735 NEW YORK AVE., N.W., WASHINGTON, DC 20006

THE AMERICAN INSTITUTE OF ARCHITECTS
FINANCIAL MANAGEMENT SYSTEM

PHASE COMPENSATION WORKSHEET

PHASE 4: DESIGN DEVELOPMENT SERVICES

Project____________________

Project #____________ Date__________

Owner____________________

Architect____________________

SERVICE	4.01	4.02	4.03	4.04	4.21	4.22	4.23	4.24	4.25	4.26	4.27	4.28	4.29	4.32	4.33		TOTAL HOURS	TOTAL DOLLARS	ITEM	LINE
	Project Administration	Disciplines Coord./ Document Checking	Agency Consulting/ Review/Approval	Owner-supplied Data Coordination	Architectural Design/ Documentation	Structural Design/ Documentation	Mechanical Design/ Documentation	Electrical Design/ Documentation	Civil Design/ Documentation	Landscape Design/ Documentation	Interior Design/ Documentation	Materials Research/ Specifications	Project Development Scheduling	Statement of Probable Construction Cost	Presentations					
IN-HOUSE PERSONNEL Hrs. / $																			@ $	1
Hrs. / $																			@ $	2
Hrs. / $																			@ $	3
Hrs. / $																			@ $	4
Hrs. / $																			@ $	5
SUB TOT. Hrs. / $																			Direct In-house Salary Expense	6
																			Direct Personnel Expense	7
																			Indirect Expense	8
																			Other Nonreimbursable Direct Expense	9
																			Total In-house Expense	10
OUT-SIDE Hrs. / $																			Outside Services Expense	11
TOT. Hrs. / $																			Estimated Total Expense	12
REMARKS																			Contingency	13
																			Profit	14
																			Proposed Compensation	15
																			Estimated Reimbursable Expense	16

THE AMERICAN INSTITUTE OF ARCHITECTS
FINANCIAL MANAGEMENT SYSTEM

PHASE COMPENSATION WORKSHEET

SERVICE		5.01	5.02	5.03	5.04	5.21	5.22	5.23	5.24	5.25	5.26	5.27	5.28	5.30	5.32	5.33	
		Project Administration	Disciplines Coord./ Document Checking	Agency Consulting/ Review/Approval	Owner-supplied Data Coordination	Architectural Design/ Documentation	Structural Design/ Documentation	Mechanical Design/ Documentation	Electrical Design/ Documentation	Civil Design/ Documentation	Landscape Design/ Documentation	Interior Design/ Documentation	Materials Research/ Specifications	Special Bidding Documents/Scheduling	Statement of Probable Construction Cost	Presentations	
IN-HOUSE PERSONNEL	Hrs.																
	$																
	Hrs.																
	$																
	Hrs.																
	$																
	Hrs.																
	$																
	Hrs.																
	$																
SUB TOT.	Hrs.																
	$																
OUT-SIDE	Hrs.																
	$																
TOT.	Hrs.																
	$																

REMARKS

PHASE 5: CONSTRUCTION DOCUMENTS SERVICES

Project________________

Project #__________ Date__________

Owner________________

Architect________________

TOTAL HOURS	TOTAL DOLLARS	ITEM	LINE
		@ $	1
		@ $	2
		@ $	3
		@ $	4
		@ $	5
		Direct In-house Salary Expense	6
		Direct Personnel Expense	7
		Indirect Expense	8
		Other Nonreimbursable Direct Expense	9
		Total In-house Expense	10
		Outside Services Expense	11
		Estimated Total Expense	12
		Contingency	13
		Profit	14
		Proposed Compensation	15
		Estimated Reimbursable Expense	16

THE AMERICAN INSTITUTE OF ARCHITECTS
FINANCIAL MANAGEMENT SYSTEM

PHASE COMPENSATION WORKSHEET

SERVICE		6.01	6.02	6.03	6.04	6.34	6.35	6.36	6.37	6.38	6.39	6.40					
		Project Administration	Disciplines Coord./ Document Checking	Agency Consulting/ Review/Approval	Owner-supplied Data Coordination	Bidding Materials	Addenda	Bidding/Negotiations	Analysis of Alter-nates/Substitutions	Special Bidding Services	Bid Evaluation	Construction Contract Agreements					
IN-HOUSE PERSONNEL	Hrs.																
	$																
	Hrs.																
	$																
	Hrs.																
	$																
	Hrs.																
	$																
	Hrs.																
	$																
SUB TOT.	Hrs.																
	$																
OUT-SIDE	Hrs.																
	$																
TOT.	Hrs.																
	$																

REMARKS

PHASE 6: BIDDING OR NEGOTIATIONS SERVICES

Project____________________

Project #____________ Date__________

Owner____________________

Architect____________________

TOTAL HOURS	TOTAL DOLLARS	ITEM	LINE
		@ $	1
		@ $	2
		@ $	3
		@ $	4
		@ $	5
		Direct In-house Salary Expense	6
		Direct Personnel Expense	7
		Indirect Expense	8
		Other Nonreimbursable Direct Expense	9
		Total In-house Expense	10
		Outside Services Expense	11
		Estimated Total Expense	12
		Contingency	13
		Profit	14
		Proposed Compensation	15
		Estimated Reimbursable Expense	16

THE AMERICAN INSTITUTE OF ARCHITECTS
FINANCIAL MANAGEMENT SYSTEM

PHASE COMPENSATION WORKSHEET

PHASE 8: POSTCONSTRUCTION SERVICES

Project____________________

Project #______________ Date__________

Owner____________________

Architect____________________

SERVICE		8.01 Project Administration	8.02 Disciplines Coord./ Document Checking	8.03 Agency Consulting/ Review/Approval	8.04 Owner-supplied Data Coordination	8.50 Maintenance and Operational Programming	8.51 Start-up Assistance	8.52 Record Drawings	8.53 Warranty Review	8.54 Postconstruction Evaluation							
IN-HOUSE PERSONNEL	Hrs.																
	$																
	Hrs.																
	$																
	Hrs.																
	$																
	Hrs.																
	$																
	Hrs.																
	$																
SUB TOT.	Hrs.																
	$																
OUT-SIDE	Hrs.																
	$																
TOT.	Hrs.																
	$																

TOTAL HOURS	TOTAL DOLLARS	ITEM	LINE
		@ $	1
		@ $	2
		@ $	3
		@ $	4
		@ $	5
		Direct In-house Salary Expense	6
		Direct Personnel Expense	7
		Indirect Expense	8
		Other Nonreimbursable Direct Expense	9
		Total In-house Expense	10
		Outside Services Expense	11
		Estimated Total Expense	12
		Contingency	13
		Profit	14
		Proposed Compensation	15
		Estimated Reimbursable Expense	16

REMARKS

PHASE COMPENSATION WORKSHEET

PHASE 9: SUPPLEMENTAL SERVICES
(first of two pages)

Project ____________________

Project # ______________ Date ______________

Owner ____________________

Architect ____________________

SERVICE		9.61 Special Studies	9.62 Renderings	9.63 Model Construction	9.64 Life Cycle Cost Analysis	9.65 Value Analysis	9.66 Quantity Surveys	9.67 Detailed Construction Cost Estimates	9.68 Energy Studies	9.69 Environmental Monitoring	9.70 Tenant-related Services	9.71 Graphics Design	9.72 Fine Arts and Crafts Services	9.73 Special Furnishings Design	9.74 Non-Building Equipment Selection	9.75 Project Promotion/ Public Relations	SUBTOTAL
IN-HOUSE PERSONNEL	Hrs.																
	$																
	Hrs.																
	$																
	Hrs.																
	$																
	Hrs.																
	$																
	Hrs.																
	$																
SUB TOT.	Hrs.																
	$																
OUT-SIDE	Hrs.																
	$																
TOT.	Hrs.																
	$																

TOTAL HOURS	TOTAL DOLLARS	ITEM	LINE
		@ $	1
		@ $	2
		@ $	3
		@ $	4
		@ $	5
		Direct In-house Salary Expense	6
		Direct Personnel Expense	7
		Indirect Expense	8
		Other Nonreimbursable Direct Expense	9
		Total In-house Expense	10
		Outside Services Expense	11
		Estimated Total Expense	12
		Contingency	13
		Profit	14
		Proposed Compensation	15
		Estimated Reimbursable Expense	16

REMARKS

THE AMERICAN INSTITUTE OF ARCHITECTS, 1735 NEW YORK AVE., N.W., WASHINGTON, DC 20006

THE AMERICAN INSTITUTE OF ARCHITECTS
FINANCIAL MANAGEMENT SYSTEM

PHASE COMPENSATION WORKSHEET

PHASE 9: SUPPLEMENTAL SERVICES
(second of two pages)

Project________________________

Project #______________ Date______________

Owner________________________

Architect________________________

SERVICE		SUBTOTAL	9.76 Leasing Brochures	9.77 Expert Witness	9.78 Computer Applications	9.79 Materials and Systems Testing	9.80 Demolition Services	9.81 Mock-up Services	9.82 Still Photography	9.83 Motion Pictures and Videotape	9.84 Coord. with Non-Design Professionals	9.85 Special Disciplines Consultation	9.86 Special Building Type Consultation					TOTAL HOURS	TOTAL DOLLARS	ITEM	LINE
IN-HOUSE PERSONNEL	Hrs. / $																			@ $	1
	Hrs. / $																			@ $	2
	Hrs. / $																			@ $	3
	Hrs. / $																			@ $	4
	Hrs. / $																			@ $	5
SUB TOT.	Hrs. / $																			Direct In-house Salary Expense	6
																				Direct Personnel Expense	7
																				Indirect Expense	8
																				Other Nonreimbursable Direct Expense	9
																				Total In-house Expense	10
OUT-SIDE	Hrs. / $																			Outside Services Expense	11
TOT.	Hrs. / $																			Estimated Total Expense	12
REMARKS																				Contingency	13
																				Profit	14
																				Proposed Compensation	15
																				Estimated Reimbursable Expense	16

THE AMERICAN INSTITUTE OF ARCHITECTS, 1735 NEW YORK AVE., N.W., WASHINGTON, DC 20006

THE AMERICAN INSTITUTE OF ARCHITECTS
FINANCIAL MANAGEMENT SYSTEM

PROJECT COMPENSATION WORKSHEET

Project ______________________

Project # ______________ Date __________

Owner ______________________

Architect ______________________

PHASE											TOTAL HOURS	TOTAL DOLLARS	ITEM	LINE
IN-HOUSE PERSONNEL	Hrs.												@ $	1
	$													
	Hrs.												@ $	2
	$													
	Hrs.												@ $	3
	$													
	Hrs.												@ $	4
	$													
	Hrs.												@ $	5
	$													
SUB TOT.	Hrs.												Direct In-house Salary Expense	6
	$													
	$												Direct Personnel Expense	7
	$												Indirect Expense	8
	$												Other Nonreimbursable Direct Expense	9
	$												Total In-house Expense	10
OUT-SIDE	Hrs.												Outside Services Expense	11
	$													
TOT.	Hrs.												Estimated Total Expense	12
	$													
	$												Contingency	13
	$												Profit	14
	$												Proposed Compensation	15
	$												Estimated Reimbursable Expense	16

THE AMERICAN INSTITUTE OF ARCHITECTS
FINANCIAL MANAGEMENT SYSTEM

PROJECT COMPENSATION SUMMARY

Project____________________

Project #__________ Date__________

Owner____________________

Architect____________________

PHASE	Proposed Compensation Amount By Phase	Cumulative Proposed Compensation Amount	Estimated Reimbursable Expense	Date	Revision		
TOTALS							

REMARKS

THE AMERICAN INSTITUTE OF ARCHITECTS
FINANCIAL MANAGEMENT SYSTEM

PROJECT TIME AND PAYMENT SCHEDULE

Project ______________________

Project # ____________ Date ____________

Owner ______________________

Architect ______________________

	AMOUNT
PHASE	DATE
1. Predesign Services	
2. Site Analysis Services	
3. Schematic Design Services	
4. Design Development Services	
5. Construction Documents Services	
6. Bidding of Negotiations Services	
7. Construction Contract Administration Services	
8. Postconstruction Services	
9. Supplemental Services	
TOTALS	

REMARKS

THE AMERICAN INSTITUTE OF ARCHITECTS
FINANCIAL MANAGEMENT SYSTEM

PHASE/SERVICE MATRIX

Project ______________ Project # ______________

● = Time Estimates and Records ◐ = Time Estimates Only ○ = No Time Estimates or Records

	PHASE 1: PREDESIGN SERVICES 1	PHASE 2: SITE ANALYSIS SERVICES 2	PHASE 3: SCHEMATIC DESIGN SERVICES 3	PHASE 4: DESIGN DEVELOPMENT SERVICES 4	PHASE 5: CONSTRUCTION DOCUMENTS SERVICES 5	PHASE 6: BIDDING OR NEGOTIATIONS SERVICES 6	PHASE 7: CONSTRUCTION CONTRACT ADMINISTRATION SERVICES 7	PHASE 8: POST-CONSTRUCTION SERVICES 8	PHASE 9: SUPPLEMENTAL SERVICES 9a	PHASE 9: (CONT'D) SUPPLEMENTAL SERVICES 9b	
A	.01 Project Administration ○	.01 Project Administration ○	.01 Project Administration ○	.01 Project Administration ○	.01 Project Administration ○	.01 Project Administration ○	.01 Project Administration ○	.01 Project Administration ○	.61 Special Studies ○	.79 Materials and Systems Testing ○	A
B	.02 Disciplines Coordination/Document Checking ○	.02 Disciplines Coordination/Document Checking ○	.02 Disciplines Coordination/Document Checking ○	.02 Disciplines Coordination/Document Checking ○	.02 Disciplines Coordination/Document Checking ○	.02 Disciplines Coordination/Document Checking ○	.02 Disciplines Coordination/Document Checking ○	.02 Disciplines Coordination/Document Checking ○	.62 Renderings ○	.80 Demolition Services ○	B
C	.03 Agency Consulting/Review/Approval ○	.03 Agency Consulting/Review/Approval ○	.03 Agency Consulting/Review/Approval ○	.03 Agency Consulting/Review/Approval ○	.03 Agency Consulting/Review/Approval ○	.03 Agency Consulting/Review/Approval ○	.03 Agency Consulting/Review/Approval ○	.03 Agency Consulting/Review/Approval ○	.63 Model Construction ○	.81 Mock-up Services ○	C
D	.04 Owner-supplied Data Coordination ○	.04 Owner-supplied Data Coordination ○	.04 Owner-supplied Data Coordination ○	.04 Owner-supplied Data Coordination ○	.04 Owner-supplied Data Coordination ○	.04 Owner-supplied Data Coordination ○	.04 Owner-supplied Data Coordination ○	.04 Owner-supplied Data Coordination ○	.64 Life Cycle Cost Analysis ○	.82 Still Photography ○	D
E	.05 Programming ○	.13 Site Analysis and Selection ○	.21 Archiectural Design/Documentation ○	.21 Architectural Design/Documentation ○	.21 Architectural Design/Documentation ○	.34 Bidding Materials ○	.41 Office Construction Administration ○	.50 Maintenance and Operational Programming ○	.65 Value Analysis ○	.83 Motion Pictures and Videotape ○	E
F	.06 Space Schematics/Flow Diagrams ○	.14 Site Development Planning ○	.22 Structural Design/Documentation ○	.22 Structural Design/Documentation ○	.22 Structural Design/Documentation ○	.35 Addenda ○	.42 Construction Field Observation ○	.51 Start-up Assistance ○	.66 Quantity Surveys ○	.84 Coordination with Non-Design Professionals ○	F
G	.07 Existing Facilities Surveys ○	.15 Detailed Site Utilization Studies ○	.23 Mechanical Design/Documentation ○	.23 Mechanical Design/Documentation ○	.23 Mechanical Design/Documentation ○	.36 Bidding/Negotiations ○	.43 Project Representation ○	.52 Record Drawings ○	.67 Detailed Construction Cost Estimates ○	.85 Special Disciplines Consultation ○	G
H	.08 Marketing Studies ○	.16 On-site Utility Studies ○	.24 Electrical Design/Documentation ○	.24 Electrical Design/Documentation ○	.24 Electrical Design/Documentation ○	.37 Analysis of Alternates/Substitutions ○	.44 Inspection Coordination ○	.53 Warranty Review ○	.68 Energy Studies ○	.86 Special Building Type Consultation ○	H
I	.09 Economic Feasibility Studies ○	.17 Off-site Utility Studies ○	.25 Civil Design/Documentation ○	.25 Civil Design/Documentation ○	.25 Civil Design/Documentation ○	.38 Special Bidding Services ○	.45 Supplemental Documents ○	.54 Postconstruction Evaluation ○	.69 Environmental Monitoring ○	○	I
J	.10 Project Financing ○	.18 Environmental Studies and Reports ○	.26 Landscape Design/Documentation ○	.26 Landscape Design/Documentation ○	.26 Landscape Design/Documentation ○	.39 Bid Evaluation ○	.46 Quotation Requests/Change Orders ○	○	.70 Tenant-related Services ○	○	J
K	○	.19 Zoning Processing Assistance ○	.27 Interior Design/Documentation ○	.27 Interior Design/Documentation ○	.27 Interior Design/Documentation ○	.40 Construction Contract Agreements ○	.47 Project Schedule Monitoring ○	○	.71 Graphics Design ○	○	K
L	○	○	.28 Materials Research/Specifications ○	.28 Materials Research/Specifications ○	.28 Materials Research/Specifications ○	○	.48 Construction Cost Accounting ○	○	.72 Fine Arts and Crafts Services ○	○	L
M	.29 Project Development Scheduling ○	.29 Project Development Scheduling ○	.29 Project Development Scheduling ○	.29 Project Development Scheduling ○	.30 Special Bidding Documents/Scheduling ○	○	.49 Project Closeout ○	○	.73 Special Furnishings Design ○	○	M
N	.31 Project Budgeting ○	.31 Project Budgeting ○	.32 Statement of Probable Construction Cost ○	.32 Statement of Probable Construction Cost ○	.32 Statement of Probable Construction Cost ○	○	○	○	.74 Non-Building Equipment Selection ○	○	N
O	.33 Presentations ○	.33 Presentations ○	.33 Presentations ○	.33 Presentations ○	.33 Presentations ○	○	○	○	.75 Project Promotion/Public Relations ○	○	O
P	○	○	○	○	○	○	○	○	.76 Leasing Brochures ○	○	P
Q	○	○	○	○	○	○	○	○	.77 Expert Witness ○	○	Q
R	○	○	○	○	○	○	○	○	.78 Computer Applications ○	○	R

THE AMERICAN INSTITUTE OF ARCHITECTS

AIA Document B161

Standard Form of Agreement Between Owner and Architect For Designated Services

1977 EDITION

This Document is intended to be used in conjunction with AIA Document B162, Scope of Designated Services.

THIS DOCUMENT HAS IMPORTANT LEGAL CONSEQUENCES; CONSULTATION WITH AN ATTORNEY IS ENCOURAGED WITH RESPECT TO ITS COMPLETION OR MODIFICATION

AGREEMENT

made as of the day of in the year of Nineteen Hundred and

BETWEEN the Owner:

and the Architect:

For the following Project:
(Include detailed description of Project location and scope.)

The Owner and the Architect agree as set forth below.

(This page blank)

TERMS AND CONDITIONS OF AGREEMENT BETWEEN OWNER AND ARCHITECT

ARTICLE 1

1.1 DESIGNATED SERVICES

The Architect's Designated Services consist of those services agreed to be performed directly by the Architect, through the Architect by utilization of Outside Services, and by Coordination Services performed by the Architect on services provided by the Owner, for and necessary to the Project, as identified and described in the Scope of Designated Services attached to and made a part of this Agreement.

1.2 CONSTRUCTION CONTRACT ADMINISTRATION PHASE

The following terms and conditions shall apply to the relevant Construction Contract Administration Phase services, if any, as may be included in the Scope of Designated Services attached to and made a part of this Agreement.

1.2.1 The Construction Phase will commence with the award of the Contract for Construction and, together with the Architect's obligation to provide services under this phase of the Agreement, will terminate when final payment to the Contractor is due, or in the absence of a final Certificate for Payment or of such due date, sixty days after the Date of Substantial Completion of the Work, whichever occurs first.

1.2.2 Unless otherwise provided in this Agreement and incorporated in the Contract Documents, the Architect shall provide administration of the Contract for Construction as set forth below and in the edition of AIA Document A201, General Conditions of the Contract for Construction, current as of the date of this Agreement.

1.2.3 The Architect shall be a representative of the Owner during the Construction Phase, and shall advise and consult with the Owner. Instructions to the Contractor shall be forwarded through the Architect. The Architect shall have authority to act on behalf of the Owner only to the extent provided in the Contract Documents unless otherwise modified by written instrument in accordance with Subparagraph 1.2.16.

1.2.4 The Architect shall visit the site at intervals appropriate to the stage of construction or as otherwise agreed by the Architect in writing to become generally familiar with the progress and quality of the Work and to determine in general if the Work is proceeding in accordance with the Contract Documents. However, the Architect shall not be required to make exhaustive or continuous on-site inspections to check the quality or quantity of the Work. On the basis of such on-site observations as an architect, the Architect shall keep the Owner informed of the progress and quality of the Work, and shall endeavor to guard the Owner against defects and deficiencies in the Work of the Contractor.

1.2.5 The Architect shall not have control or charge of and shall not be responsible for construction means, methods, techniques, sequences or procedures, or for safety precautions and programs in connection with the Work, for the acts or omissions of the Contractor, Subcontractors or any other persons performing any of the Work, or for the failure of any of them to carry out the Work in accordance with the Contract Documents.

1.2.6 The Architect shall at all times have access to the Work wherever it is in preparation or progress.

1.2.7 The Architect shall determine the amounts owing to the Contractor based on observations at the site and on evaluations of the Contractor's Applications for Payment, and shall issue Certificates for Payment in such amounts, as provided in the Contract Documents.

1.2.8 The issuance of a Certificate for Payment shall constitute a representation by the Architect to the Owner, based on the Architect's observations at the site as provided in Subparagraph 1.2.4 and the data comprising the Contractor's Application for Payment, that the Work has progressed to the point indicated; that, to the best of the Architect's knowledge, information and belief, the quality of the Work is in accordance with the Contract Documents (subject to an evaluation of the Work for conformance with the Contract Documents upon Substantial Completion, to the results of any subsequent tests required by or performed under the Contract Documents, to minor deviations from the Contract Documents correctable prior to completion, and to any specific qualifications stated in the Certificate for Payment); and that the Contractor is entitled to payment in the amount certified. However, the issuance of a Certificate for Payment shall not be a representation that the Architect has made any examination to ascertain how and for what purpose the Contractor has used the moneys paid on account of the Contract Sum.

1.2.9 The Architect shall be the interpreter of the requirements of the Contract Documents and the judge of the performance thereunder by both the Owner and Contractor. The Architect shall render interpretations necessary for the proper execution or progress of the Work with reasonable promptness on written request of either the Owner or the Contractor, and shall render written decisions, within a reasonable time, on all claims, disputes and other matters in question between the Owner and the Contractor relating to the execution or progress of the Work or the interpretation of the Contract Documents.

1.2.10 Interpretations and decisions of the Architect shall be consistent with the intent of and reasonably inferable from the Contract Documents and shall be in written or graphic form. In the capacity of interpreter and judge, the Architect shall endeavor to secure faithful performance by both the Owner and the Contractor, shall not show partiality to either, and shall not be liable for the result of any interpretation or decision rendered in good faith in such capacity.

1.2.11 The Architect's decisions in matters relating to artistic effect shall be final if consistent with the intent of the Contract Documents. The Architect's decisions on any other claims, disputes or other matters, including those in question between the Owner and the Contractor, shall be subject to arbitration as provided in this Agreement and in the Contract Documents.

1.2.12 The Architect shall have authority to reject Work which does not conform to the Contract Documents. Whenever, in the Architect's reasonable opinion, it is necessary or advisable for the implementation of the intent of the Contract Documents, the Architect will have authority to require special inspection or testing of the Work in accordance with the provisions of the Contract Documents, whether or not such Work be then fabricated, installed or completed.

1.2.13 The Architect shall review and approve or take other appropriate action upon the Contractor's submittals such as Shop Drawings, Product Data, and Samples, but only for conformance with the design concept of the Work and with the information given in the Contract Documents. Such action shall be taken with reasonable promptness so as to cause no delay. The Architect's approval of a specific item shall not indicate approval of an assembly of which the item is a component.

1.2.14 The Architect shall prepare Change Orders for the Owner's approval and execution in accordance with the Contract Documents, and shall have authority to order minor changes in the Work not involving an adjustment in the Contract Sum or an extension of the Contract Time which are not inconsistent with the intent of the Contract Documents.

1.2.15 The Architect shall conduct inspections to determine the Dates of Substantial Completion and final completion, shall receive and forward to the Owner for the Owner's review written warranties and related documents required by the Contract Documents and assembled by the Contractor, and shall issue a final Certificate for Payment.

1.2.16 The extent of the duties, responsibilities, and limitations of authority of the Architect as the Owner's representative during construction shall not be modified or extended without written consent of the Owner, the Contractor and the Architect.

1.3 PROJECT REPRESENTATION BEYOND DESIGNATED SERVICES

1.3.1 If the Owner and Architect agree that more extensive representation at the site than is described in Paragraph 1.2 shall be provided, the Architect shall provide one or more Project Representatives to assist the Architect in carrying out such responsibilities at the site.

1.3.2 Such Project Representatives shall be selected, employed and directed by the Architect, and the Architect shall be compensated therefor as mutually agreed between the Owner and the Architect as set forth in an exhibit appended to this Agreement, which shall describe the duties, responsibilities and limitations of authority of such Project Representatives.

1.3.3 Through the observations by such Project Representatives, the Architect shall endeavor to provide further protection for the Owner against defects and deficiencies in the Work, but the furnishing of such project representation shall not modify the rights, responsibilities or obligations of the Architect as described in Paragraph 1.2.

1.4 ADDITIONAL SERVICES

The following Services are not included as Designated Services unless so identified in Article 15. These Services and any other services identified as Additional Services in the Scope of Designated Services attached to and made a part of this Agreement shall be provided if authorized or confirmed in writing by the Owner, and they shall be paid for by the Owner as provided in this Agreement, in addition to the compensation for Designated Services.

1.4.1 Providing services in connection with the work of a construction manager or separate consultants retained by the Owner.

1.4.2 Making revisions in Drawings, Specifications or other documents when such revisions are inconsistent with written approvals or instructions previously given, are required by the enactment or revision of codes, laws or regulations subsequent to the preparation of such documents or are due to other causes not solely within the control of the Architect.

1.4.3 Preparing Drawings, Specifications, supporting data and providing other services in connection with Change Orders to the extent that the adjustment in the Basic Compensation resulting from the adjusted Construction Cost is not commensurate with the services required of the Architect, provided such Change Orders are required by causes not solely within the control of the Architect.

1.4.4 Providing consultation concerning replacement of any Work damaged by fire or other cause during construction, and furnishing services as may be required in connection with the replacement of such Work.

1.4.5 Providing services made necessary by the default of the Contractor, or by major defects or deficiencies in the Work of the Contractor, or by failure of performance of either the Owner or Contractor under the Contract for Construction.

1.4.6 Providing services, other than those identified in the Scope of Designated Services, after issuance to the Owner of the final Certificate for Payment, or in the absence of a final Certificate for Payment, more than sixty days after the Date of Substantial Completion of the Work.

1.4.7 Providing any other services not otherwise included in this Agreement or not customarily furnished in accordance with generally accepted architectural practice.

1.5 TIME

1.5.1 The Architect shall perform Designated and Additional Services as expeditiously as is consistent with professional skill and care and the orderly progress of the Work. Upon request of the Owner, the Architect shall submit for the Owner's approval a schedule for the performance of the Architect's services which shall be adjusted as required as the Project proceeds, and shall include allowances for periods of time required for the Owner's review and approval of submissions and for approvals of authorities having jurisdiction over the Project. This schedule, when approved by the Owner, shall not, except for reasonable cause, be exceeded by the Architect.

ARTICLE 2
THE OWNER'S RESPONSIBILITIES

The following responsibilities will be undertaken by the Owner, in addition to those Services designated to be performed by the Owner and described in the Scope of Designated Services attached to and made a part of this Agreement.

2.1 The Owner shall provide full information regarding requirements for the Project.

2.2 If the Owner provides a budget for the Project it shall include contingencies for bidding, changes in the Work during construction, and other costs which are the responsibility of the Owner, including those described in this Article 2 and in Subparagraph 3.1.2. The Owner shall, at the request of the Architect, provide a statement of funds available for the Project, and their source.

2.3 The Owner shall designate, when necessary, a representative authorized to act in the Owner's behalf with respect to the Project. The Owner or such authorized representative shall examine the documents submitted by the Architect and shall render decisions pertaining thereto promptly, to avoid unreasonable delay in the progress of the Architect's services.

2.4 The Owner shall furnish a legal description and a certified land survey of the site, giving, as applicable, grades and lines of streets, alleys, pavements and adjoining property; rights-of-way, restrictions, easements, encroachments, zoning, deed restrictions, boundaries and contours of the site; locations, dimensions and complete data pertaining to existing buildings, other improvements and trees; and full information concerning available service and utility lines both public and private, above and below grade, including inverts and depths.

2.5 The Owner shall furnish the services of soil engineers or other consultants when such services are deemed necessary by the Architect. Such services shall include test borings, test pits, soil bearing values, percolation tests, air and water pollution tests, ground corrosion and resistivity tests, including necessary operations for determining subsoil, air and water conditions, with reports and appropriate professional recommendations.

2.6 The Owner shall furnish structural, mechanical, chemical and other laboratory tests, inspections and reports as required by law or the Contract Documents.

2.7 The Owner shall furnish all legal, accounting and insurance counseling services as may be necessary at any time for the Project, including such auditing services as the Owner may require to verify the Contractor's Applications for Payment or to ascertain how or for what purposes the Contractor uses the moneys paid by or on behalf of the Owner.

2.8 The services, information, surveys and reports required by Paragraphs 2.4 through 2.7 inclusive shall be furnished at the Owner's expense, and the Architect shall be entitled to rely upon the accuracy and completeness thereof.

2.9 If the Owner observes or otherwise becomes aware of any fault or defect in the Project or nonconformance with the Contract Documents, prompt written notice thereof shall be given by the Owner to the Architect.

2.10 The Owner shall furnish required information and services and shall render approvals and decisions as expeditiously as necessary for the orderly progress of the Architect's services and of the Work.

ARTICLE 3
CONSTRUCTION COST

3.1 DEFINITION

3.1.1 The Construction Cost shall be the total cost or estimated cost to the Owner of all elements of the Project designed or specified by the Architect.

3.1.2 The Construction Cost shall include at current market rates, including a reasonable allowance for overhead and profit, the cost of labor and materials furnished by the Owner and any equipment which has been designed, specified, selected or specially provided for by the Architect.

3.1.3 Construction Cost does not include the compensation of the Architect and the Architect's consultants, the cost of the land, rights-of-way, or other costs which are the responsibility of the Owner as provided in Article 2.

3.2 RESPONSIBILITY FOR CONSTRUCTION COST

3.2.1 Evaluations of the Owner's Project budget, Statements of Probable Construction Cost and Detailed Estimates of Construction Cost, if any, prepared by the Architect, represent the Architect's best judgment as a design professional familiar with the construction industry. It is recognized, however, that neither the Architect nor the Owner has control over the cost of labor, materials or equipment, over the Contractor's methods of determining bid prices, or over competitive bidding, market or negotiating conditions. Accordingly, the Architect cannot and does not warrant or represent that bids or negotiated prices will not vary from the Project budget proposed, established or approved by the Owner, if any, or from any Statement of Probable Construction Cost or other cost estimate or evaluation prepared by the Architect.

3.2.2 No fixed limit of Construction Cost shall be established as a condition of this Agreement by the furnishing, proposal or establishment of a Project budget under Paragraph 2.2 or otherwise, unless such fixed limit has been agreed upon in writing and signed by the parties hereto. If such a fixed limit has been established, the Architect shall be permitted to include contingencies for design, bidding and price escalation, to determine what materials, equipment, component systems and types of construction are to be included in the Contract Documents, to make reasonable adjustments in the scope of the Project and to include in the Contract Documents alternate bids to adjust the Construction Cost to the fixed limit. Any such fixed limit shall be increased in the amount of any increase in the Contract Sum occurring after execution of the Contract for Construction.

3.2.3 If the Bidding or Negotiations Phase has not commenced within three months after the Architect submits the Construction Documents to the Owner, any Project budget or fixed limit of Construction Cost shall be adjusted to reflect any change in the general level of prices in the construction industry between the date of submission of the Construction Documents to the Owner and the date on which proposals are sought.

3.2.4 If a Project budget or fixed limit of Construction Cost (adjusted as provided in Subparagraph 3.2.3) is exceeded by the lowest bona fide bid or negotiated proposal, the Owner shall (1) give written approval of an increase in such fixed limit, (2) authorize rebidding or renegotiating of the Project within a reasonable time, (3) if the Project is abandoned, terminate in accordance with Paragraph 10.2, or (4) cooperate in revising the Project scope and quality as required to reduce the Construction Cost. In the case of (4), provided a fixed limit of Construction Cost has been established as a condition of this Agreement, the Architect, without additional charge, shall modify the Drawings and Specifications as necessary to comply with the fixed limit. The providing of such service shall be the limit of the Architect's responsibility arising from the establishment of such fixed limit, and having done so, the Architect shall be entitled to compensation for all services performed, in accordance with this Agreement, whether or not the Construction Phase is commenced.

ARTICLE 4
DIRECT SALARY AND DIRECT PERSONNEL EXPENSE

4.1 Direct Salary Expense is defined as the direct salaries of all the Architect's personnel engaged on the Project, but does not include the cost of contributions and benefits related thereto, whether mandatory or customary, as described in Paragraph 4.2 and included in Direct Personnel Expense.

4.2 Direct Personnel Expense is defined as the direct salaries of all the Architect's personnel engaged on the Project, and the portion of the cost of their mandatory and customary contributions and benefits related thereto, such as employment taxes and other statutory employee benefits, insurance, sick leave, holidays, vacations, pensions and similar contributions and benefits.

ARTICLE 5
REIMBURSABLE EXPENSES

5.1 Reimbursable Expenses are in addition to the Compensation for Designated and Additional Services and include actual expenditures made by the Architect and the Architect's employees and consultants in the interest of the Project for the expenses listed in the following Subparagraphs:

5.1.1 Expense of transportation in connection with the Project; living expenses in connection with out-of-town travel; long distance communications, and fees paid for securing approval of authorities having jurisdiction over the Project.

5.1.2 Expense of reproductions, postage and handling of Drawings, Specifications and other documents, excluding reproductions for the office use of the Architect and the Architect's consultants.

5.1.3 Expense of data processing and photographic production techniques when used in connection with Additional Services.

5.1.4 If authorized in advance by the Owner, expense of overtime work requiring higher than regular rates.

5.1.5 Expense of any additional insurance coverage or limits, including professional liability insurance, requested by the Owner in excess of that normally carried by the Architect and the Architect's consultants.

ARTICLE 6
PAYMENTS TO THE ARCHITECT

6.1 PAYMENTS ON ACCOUNT OF DESIGNATED SERVICES

6.1.1 An initial payment as set forth in Paragraph 14.1 is the minimum payment under this Agreement.

6.1.2 Subsequent payments for Designated Services shall be made monthly and shall be in proportion to services performed within each Phase of Services, on the basis set forth in Article 14.

6.1.3 If and to the extent that the Contract Time initially established in the Contract for Construction is exceeded or extended through no fault of the Architect, compensation for any Designated Services required for such extended period of Administration of the Construction Contract shall be computed as set forth in Paragraph 14.4 for Additional Services.

6.1.4 When compensation is based on a percentage of Construction Cost, and any portions of the Project are deleted or otherwise not constructed, compensation for such portions of the Project shall be payable to the extent services are performed on such portions, in accordance with the schedule set forth in Subparagraph 14.2.2, based on (1) the lowest bona fide bid or negotiated proposal or, (2) if no such bid or proposal is received, the most recent Statement of Probable Construction Cost or Detailed Estimate of Construction Cost for such portions of the Project.

6.2 PAYMENTS ON ACCOUNT OF ADDITIONAL SERVICES

6.2.1 Payments on account of the Architect's Additional Services as defined in Paragraph 1.4 and for Reimbursable Expenses as defined in Article 5 shall be made monthly upon presentation of the Architect's statement of services rendered or expenses incurred.

6.3 PAYMENTS WITHHELD

6.3.1 No deductions shall be made from the Architect's compensation on account of penalty, liquidated damages or other sums withheld from payments to contractors, or on account of the cost of changes in the Work other than those for which the Architect is held legally liable.

6.4 PROJECT SUSPENSION OR TERMINATION

6.4.1 If the Project is suspended or abandoned in whole or in part for more than three months the Architect shall be compensated for all services performed prior to receipt of written notice from the Owner of such suspension or abandonment, together with Reimbursable Expenses then due and all Termination Expenses as defined in Paragraph 10.4. If the Project is resumed after being suspended for more than three months, the Architect's compensation shall be equitably adjusted.

ARTICLE 7
ARCHITECT'S ACCOUNTING RECORDS

7.1 Records of Reimbursable Expenses and expenses pertaining to Additional Services and services performed on the basis of a Multiple of Direct Personnel Expense shall be kept on the basis of generally accepted accounting principles and shall be available to the Owner or the Owner's authorized representative at mutually convenient times.

ARTICLE 8
OWNERSHIP AND USE OF DOCUMENTS

8.1 Drawings and Specifications as instruments of service are and shall remain the property of the Architect whether the Project for which they are made is executed or not. The Owner shall be permitted to retain copies, including reproducible copies, of Drawings and Specifications for information and reference in connection with the Owner's use and occupancy of the Project. The Drawings and Specifications shall not be used by the Owner on other projects, for additions to this Project or for completion of this Project by others provided the Architect is not in default under this Agreement, except by agreement in writing and with appropriate compensation to the Architect.

8.2 Submission or distribution to meet official regulatory requirements or for other purposes in connection with the Project is not to be construed as publication in derogation of the Architect's rights.

ARTICLE 9
ARBITRATION

9.1 All claims, disputes and other matters in question between the parties to this Agreement, arising out of or relating to this Agreement or the breach thereof, shall be decided by arbitration in accordance with the Construction Industry Arbitration Rules of the American Arbitration Association then obtaining unless the parties mutually agree otherwise. No arbitration, arising out of or relating to this Agreement, shall include, by consolidation, joinder or in any other manner, any additional person not a party to this Agreement except by written consent containing a specific reference to this Agreement and signed by the Architect, the Owner, and any other person sought to be joined. Any consent to arbitration involving an additional person or persons shall not constitute consent to arbitration of any dispute not described therein or with any person not named or described therein. This Agreement to arbitrate and any agreement to arbitrate with an additional person or persons duly consented to by the parties to this Agreement shall be specifically enforceable under the prevailing arbitration law.

9.2 Notice of the demand for arbitration shall be filed in writing with the other party to this Agreement and with the American Arbitration Association. The demand shall be made within a reasonable time after the claim, dispute or other matter in question has arisen. In no event shall the demand for arbitration be made after the date when institution of legal or equitable proceedings based on such claim, dispute or other matter in question would be barred by the applicable statute of limitations.

9.3 The award rendered by the arbitrators shall be final, and judgment may be entered upon it in accordance with applicable law in any court having jurisdiction thereof.

ARTICLE 10
TERMINATION OF AGREEMENT

10.1 This Agreement may be terminated by either party upon seven days' written notice should the other party fail substantially to perform in accordance with its terms through no fault of the party initiating the termination.

10.2 This Agreement may be terminated by the Owner upon at least seven days' written notice to the Architect in the event that the Project is permanently abandoned.

10.3 In the event of termination not the fault of the Architect, the Architect shall be compensated for all services performed to termination date, together with Reimbursable Expenses then due and all Termination Expenses as defined in Paragraph 10.4.

10.4 Termination Expenses include expenses directly attributable to termination for which the Architect is not otherwise compensated, plus an amount computed as a percentage of the total Compensation for Designated and Additional Services earned to the time of termination, as follows:

- **.1** 20 percent if termination occurs during the Predesign, Site Analysis or Schematic Design Phase; or
- **.2** 10 percent if termination occurs during the Design Development Phase; or
- **.3** 5 percent if termination occurs during any subsequent phase.

ARTICLE 11
MISCELLANEOUS PROVISIONS

11.1 Unless otherwise specified, this Agreement shall be governed by the law of the principal place of business of the Architect.

11.2 Terms in this Agreement shall have the same meaning as those in AIA Document A201, General Conditions of the Contract for Construction, current as of the date of this Agreement.

11.3 As between the parties to this Agreement: as to all acts or failures to act by either party to this Agreement, any applicable statute of limitations shall commence to run and any alleged cause of action shall be deemed to have accrued in any and all events not later than the relevant Date of Substantial Completion of the Work, and as to any acts or failures to act occurring after the relevant Date of Substantial Completion, not later than the date of issuance of the final Certificate for Payment.

11.4 The Owner and the Architect waive all rights against each other and against the contractors, consultants, agents and employees of the other for damages covered by any property insurance during construction as set forth in the edition of AIA Document A201, General Conditions, current as of the date of this Agreement. The Owner and the Architect each shall require appropriate similar waivers from their contractors, consultants and agents.

ARTICLE 12
SUCCESSORS AND ASSIGNS

12.1 The Owner and the Architect, respectively, bind themselves, their partners, successors, assigns and legal representatives to the other party to this Agreement and to the partners, successors, assigns and legal representatives of such other party with respect to all covenants of this Agreement. Neither the Owner nor the Architect shall assign, sublet or transfer any interest in this Agreement without the written consent of the other.

ARTICLE 13
EXTENT OF AGREEMENT

13.1 This Agreement represents the entire and integrated agreement between the Owner and the Architect and supersedes all prior negotiations, representations or agreements, either written or oral. This Agreement may be amended only by written instrument signed by both Owner and Architect.

ARTICLE 14

BASIS OF COMPENSATION

The Owner shall compensate the Architect for the Scope of Services provided, in accordance with Article 6, Payments to the Architect, and the other Terms and Conditions of this Agreement, as follows:

14.1 AN INITIAL PAYMENT of dollars ($) shall be made upon execution of this Agreement and credited to the Owner's account as follows:

14.2 COMPENSATION FOR DESIGNATED SERVICES

14.2.1 FOR DESIGNATED SERVICES, as identified and described in the Scope of Designated Services attached to and made a part of this Agreement, and any other services included in Article 15 as part of Designated Services, compensation shall be computed by the following methods for the particular Services so indicated in the Scope of Designated Services:

(Here identify basis of compensation, including fixed amounts, rates, multiples or percentages, as applicable, and identify Phases to which particular methods of compensation apply, as appropriate.)

METHOD A: MULTIPLE OF DIRECT SALARY EXPENSE

Compensation for services rendered by Principals and employees shall be based on a multiple of ( ) times their Direct Salary Expense as defined in Paragraph 4.1, and for consultants as described in Subparagraph 14.4.2.

METHOD B: MULTIPLE OF DIRECT PERSONNEL EXPENSE

Compensation for services rendered by Principals and employees shall be based on a multiple of ( ) times their Direct Personnel Expense as defined in Paragraph 4.2, and for consultants as described in Subparagraph 14.4.2.

METHOD C: PROFESSIONAL FEE PLUS EXPENSES

Compensation shall be based on a Professional Fee of dollars ($) plus compensation for services rendered by Principals and employees based on a multiple of ( ) times their Direct (Salary) (Personnel) Expense as defined in Article 4, and for consultants as described in Subparagraph 14.4.2.

METHOD D: PERCENTAGE OF CONSTRUCTION COST

Compensation shall be based on percent (%) of the Construction Cost as defined in Article 3.

METHOD E: STIPULATED SUM

Compensation shall be a Stipulated Sum of dollars ($).

METHOD F: HOURLY BILLING RATES

Compensation for services rendered by Principals and employees shall be based on the following hourly rates, and for consultants as described in Subparagraph 14.4.2:

1. Principals' time at the fixed rate of dollars ($) per hour. For the purposes of this Article, the Principals are:

2. Supervisory time at the fixed rate of dollars ($) per hour. For the purposes of this Article, Supervisory personnel include those in the following positions:

3. Technical Level I time at the fixed rate of dollars ($) per hour. For the purposes of this Article, Technical Level I personnel include those in the following positions:

4. Technical Level II time at the fixed rate of dollars ($) per hour. For the purposes of this Article, Technical Level II personnel include those in the following positions:

5. Technical Level III time at the fixed rate of dollars ($) per hour. For the purposes of this Article, Technical Level III personnel include those in the following positions:

METHOD G: MULTIPLE OF AMOUNTS BILLED TO ARCHITECT

Services of consultants at a multiple of ( ) times the amounts billed to the Architect for such services.

METHOD H:

Compensation shall be computed as follows:

14.2.2 Where compensation is based on a Stipulated Sum or Percentage of Construction Cost, payments for Designated Services shall be made as provided in Subparagraph 6.1.2, so that Compensation for each Phase to which such Method applies shall equal the following percentages of the total Compensation determined by such Method:

1. Predesign Phase: percent (%)
2. Site Analysis Phase: percent (%)
3. Schematic Design Phase: percent (%)
4. Design Development Phase: percent (%)
5. Construction Documents Phase: percent (%)
6. Bidding or Negotiation Phase: percent (%)
7. Construction Contract Administration Phase: percent (%)
8. Postconstruction Phase: percent (%)
9. Supplemental Services: percent (%)

14.3 FOR PROJECT REPRESENTATION BEYOND DESIGNATED SERVICES, as described in Paragraph 1.3, Compensation shall be computed separately in accordance with Subparagraph 1.3.2.

14.4 COMPENSATION FOR ADDITIONAL SERVICES

14.4.1 FOR ADDITIONAL SERVICES OF THE ARCHITECT, as described in Paragraph 1.4 and identified in the Scope of Designated Services as Additional Services, but excluding Additional Services of consultants, compensation shall be computed based on Method as described in Subparagraph 14.2.1.

(Above insert the identifying letter and name of method of compensation applicable to Additional Services of the Architect.)

14.4.2 FOR ADDITIONAL SERVICES OF CONSULTANTS, including additional structural, mechanical and electrical engineering services and those identified in Article 15 or in the Scope of Designated Services attached to and made a part of this Agreement, as part of Additional Services, a multiple of ( ) times the amounts billed to the Architect for such services.

14.5 FOR REIMBURSABLE EXPENSES, as described in Article 5, and any other items included in Article 15 as Reimbursable Expenses, a multiple of ( ) times the amounts expended by the Architect, the Architect's employees and consultants in the interest of the Project.

14.6 Payments due the Architect and unpaid under this Agreement shall bear interest from the date payment is due at the rate entered below, or in the absence thereof, at the legal rate prevailing at the principal place of business of the Architect.

(Here insert any rate of interest agreed upon.)

(Usury laws and requirements under the Federal Truth in Lending Act, similar state and local consumer credit laws and other regulations at the Owner's and Architect's principal places of business, the location of the Project and elsewhere may affect the validity of this provision. Specific legal advice should be obtained with respect to deletion, modification, or other requirements such as written disclosures or waivers.)

14.7 The Owner and the Architect agree in accordance with the Terms and Conditions of this Agreement that:

14.7.1 IF THE SCOPE of the Project or of the Architect's Services is changed materially, the amounts of compensation shall be equitably adjusted.

14.7.2 IF THE SERVICES covered by this Agreement have not been completed within ( ) months of the date hereof, through no fault of the Architect, the amounts of compensation, rates and multiples set forth herein shall be equitably adjusted.

ARTICLE 15

OTHER CONDITIONS OR SERVICES

This agreement entered into as of the day and year first written above.

OWNER	ARCHITECT
______________________	______________________
______________________	______________________
______________________	______________________
______________________	______________________

INSTRUCTION SHEET *AIA DOCUMENT B161a*

FOR AIA DOCUMENT B161, STANDARD FORM OF AGREEMENT BETWEEN OWNER AND ARCHITECT FOR DESIGNATED SERVICES—NOVEMBER 1977 EDITION

A. GENERAL INFORMATION:

1. New Documents B161 and B162:

This is the first edition of *AIA Document B161, Owner-Architect Agreement for Designated Services,* and is intended to be used in conjunction with *AIA Document B162, Scope of Designated Services.* These new forms are designed to work together in determining (1) the terms and conditions and the amounts of compensation of the Agreement (B161) and (2) the responsibilities and services to be undertaken by Owner and Architect (B162). The separation of B161 and B162 provides for the flexibility of using B162, the *Scope of Designated Services,* with other agreement forms. Alternatively, B161 may be used as the terms and conditions with other forms of scope of services.

In order to avoid problems of consistency and coordination, especially with regard to the methods of compensation provided for in B161 and applied to specific services in B162, B161 and B162 should be used together whenever possible. Neither B161 nor B162 may be used alone. A complete agreement will require (1) both B161 and B162, (2) B161 with another equivalent to B162, or (3) B162 with another equivalent to B161.

2. AIA Financial Management System:

Documents B161 and B162 are designed to interface with the AIA cost-based approach to compensation as set forth in *Compensation Guidelines for Architectural and Engineering Services* (AIA Catalog #M-188, 1978 Edition), and with all elements of the AIA Financial Management System. These publications and practice aids, particularly *Compensation Guidelines,* should be consulted for a full understanding of the process of establishing the scope of services and compensation.

The B161-B162 combination represents a departure from previous AIA Owner-Architect Agreements in that (1) the traditional five phases of services have been expanded to nine categories, enabling predesign, site analysis, postconstruction and supplemental services to be provided under the same terms and conditions as are other services, (2) the traditional distinction between "Basic" and "Additional" services and consequent variations in compensation methods is not recognized in these new documents, and (3) most of the services that could be provided by the Architect are included in B162, *but appropriate editing and modification by the parties prior to execution of the Agreement is required.*

3. Description:

B161 consists of the following elements:

(a) Cover page
(b) Terms and Conditions (Article 1 through 13)
(c) Compensation (Article 14)
(d) Other Conditions or Services (Article 15)

B162 consists of the following elements:

(a) Cover page
(b) Nine Phases of Services and one Phase in blank, including:
 (i) Schedules of Designated Services (grid sheets)
 (ii) Descriptions of Designated Services
(c) Enumeration of Modifications page

4. Other AIA Agreement Forms:

The following other AIA Owner-Architect Agreements are available for use in connection with traditional services or special circumstances. These should be carefully studied and used where appropriate. Users should consult AIA to determine the most recent edition of each document.

B141	Owner-Architect Agreement
B141/CM	Owner-Architect Agreement, Construction Management Edition
B151	Abbreviated Form Owner-Architect Agreement
B171	Interior Design Services Agreement
B727	Owner-Architect Agreement for Special Services
B801	Owner-Construction Manager Agreement

5. Letter Forms of Agreement:

Letter forms of agreement are generally discouraged by AIA, and users should consult the Instruction Sheet for AIA Document B141 (AIA Document B141a) and Chapter 9 of the *Architect's Handbook of Professional Practice* for further general information.

AIA Documents B161 and B162 may be used in conjunction with letter forms of agreement to establish either the terms and conditions (B161) or the scope of services (B162, as edited). Because the proper incorporation of such documents into a letter form of agreement is critical for the establishment of the intended agreement, an attorney should be consulted where B161 and B162 are not used together or do not represent the entire agreement.

6. Owner-Furnished Agreement Forms:

Many clients, especially governmental agencies, require the use of an Owner-Architect Agreement prepared by them. Because of the flexibility of the Designated Services approach (cost-based compensation), some clients may employ documents similar to B161 and B162 in form and content. Such forms should be carefully compared with AIA Documents B161 and B162 before any agreement is entered into. If there are any significant omissions, additions or variances from the terms of B161 or B162, both legal counsel and liability insurance counsel should be consulted. In numerous instances, AIA and its components have succeeded in encouraging clients to adopt the AIA Owner-Architect Agreements or to bring their own forms into better conformance with accepted architectural practice.

7. Modifications:

As with all AIA Documents, users are encouraged to consult an attorney with respect to completing the forms and with respect to any modifications that are proposed. Of particular concern is the need for consistency between the Construction Phase services and responsibilities set forth in the Owner-Architect Agreement and those set forth in the anticipated General Conditions of the Contract for Construction. Generally, modifications to B161 can be accomplished by striking out portions which are inapplicable, and inserting any additional provisions in Article 15. Legal counsel should also be sought concerning the effect of state and local law on the terms of the Agreement, particularly with respect to registration laws, duties imposed by building codes, interest charges, and arbitration.

CAUTION: Modifications *must* be made to the descriptions of services in B162 before it can be incorporated into the agreement. B162 is intended to set forth a large number of services which may potentially be provided, and must be "tailored" to the requirements of the particular project.

B. COMPLETING THE B161 AGREEMENT FORM:

1. Cover Page:

Date: The date represents the first date as of which the Agreement is entered into. It may be the date that an oral agreement was reached, the date the Agreement is submitted to the Owner, the date authorizing action was taken by the Owner, or the date of actual execution. No professional services under this Agreement should be performed prior to the date indicated.

Identification of Parties: Parties to this Agreement should be identified in the capacity in which the Agreement is to be executed, including the name of the firms and capacity of persons signing, the address of the principal office and a designation of the legal status of both parties: sole proprietorship, partnership, joint venture, unincorporated association, limited partnership or corporation (general, close, or professional), etc. Where appropriate a copy of the resolution authorizing the individual to act on behalf of the firm or entity should be attached.

Project Description: The proposed project should be described in sufficient detail to identify (1) the official name or title of the facility, (2) the location of the site, if known, (3) the proposed building type or usage, (4) the usage or purpose to which the facility will be put, (5) the size or capacity of the facility. The construction cost (whether a fixed limit or Project budget) should be indicated in Article 15.

2. Article 1—DESIGNATED SERVICES:

Paragraph 1.2: Note that B161 provides descriptions of the Architect's Construction Phase Services which are coordinated with AIA Document A201, General Conditions of the Contract for Construction. The identification of the services to be provided during construction will occur in B162. If any of the services described in Paragraph 1.2 of B161 will not be included in the services designated in B162, they should be stricken from Paragraph 1.2. Modifications to these descriptions should be made in Article 15.

3. Article 2—THE OWNER'S RESPONSIBILITIES:

Because B162 provides for services to be undertaken by the Owner as well as the Architect, any services described in this Article should be modified as appropriate or supplemented by making reference to B162, as described above for the Architect's services in Paragraph 1.2.

4. Article 14—BASIS OF COMPENSATION:

Paragraph 14.1: Insert the amount of the retainer, if any, and indicate whether it will be credited to the first, the last, or proportionately to all of the payments on the Owner's account.

Paragraph 14.2, Subparagraph 14.2.1: The Agreement form (B161) provides for the multiple methods of compensation (A through H) which are applied to specific services in B162, Scope of Designated Services. These Methods are described in *Compensation Guidelines.* On each of the B162 "grid sheets," the right-hand column is used to identify the method of compensation applicable for each selected Service. Once the methods of compensation to be used on the Project have been identified in B162, Subparagraph 14.2.1 in B161 should be completed by filling in the blanks of the applicable methods, and striking out those which will not be used. If a method applies only to particular Phases of service, this may be indicated following the title of the compensation method in 14.2.1.

Paragraph 14.2, Subparagraph 14.2.2: Insert the appropriate percentages, words and numbers, deleting inapplicable Phases. Strike out Paragraph if appropriate.

Paragraph 14.4, Subparagraph 14.4.1: Insert the appropriate basis for compensation, referring to the methods provided for in Subparagraph 14.2.1 as a guide. Most frequently used methods of compensation for Additional Services are Methods A, B or F.

Paragraph 14.4, Subparagraph 14.4.2: Insert the multiplier applied to consultant billings used to cover the costs of administration, responsibility for consultants' work, coordination and profit. If a mark-up will not be used, insert "one (1.0)" as the multiple.

Paragraph 14.5: Insert the multiplier applied to reimbursable expenses used to cover the costs of administration. If a mark-up will not be used, insert "one (1.0)" as the multiple.

Paragraph 14.6: Insert the percentage rate and basis (monthly, annual) and the time (such as a number of days) after the due date for payment on which interest charges will begin to run. This should be carefully checked against state usury laws which may set a limit on the rate of interest which may legally be charged. In addition, Federal Truth in Lending and similar state and local consumer protection laws may require setting forth the annual percentage rate and other disclosures or waivers for certain types of transactions or with certain types of clients. Advice of legal counsel should be sought on such matters.

Paragraph 14.7: Insert the amount of time after which the compensation shall be subject to renegotiation or adjustment. If the firm requires periodic adjustments in hourly rates and multiples, this should be stated, along with limitations, if any, on the amount of upward adjustment which may be made.

5. **Article 15—OTHER CONDITIONS OR SERVICES:** Here insert the following types of provisions:
   - (a) Designation of which Phases of services are included in, and which are excluded from, the Agreement.
   - (b) General Conditions to be utilized if different from AIA Document A201.
   - (c) Modifications to any conditions (other than modifications of service descriptions which are made in B162).
   - (d) Other additional conditions (other than modifications of service descriptions which are made in B162).
   - (e) Identification of consultants to parties to this Agreement.
   - (f) Identification of other entities performing services for the Project under special arrangements with a party to this Agreement.
   - (g) Construction cost (fixed limit or Project budget), if any.
   - (h) Construction delivery process anticipated to be used (single, separate, phased contracts, etc.).
   - (i) Time schedule for performance of services, if any.

   Note comments on Construction Phase services under Article 1 above.

6. **Execution of the Agreement:**
   In executing the Agreement, the parties should indicate the capacity in which they are acting.

C. SCOPE OF SERVICES:

See the Instructions for B162 (AIA Document B162a) for a discussion of the Scope of Designated Services.

D. REPRODUCTION:

THE AMERICAN INSTITUTE OF ARCHITECTS

AIA Document B162

Scope of Designated Services

1977 EDITION

THIS DOCUMENT HAS IMPORTANT LEGAL CONSEQUENCES; CONSULTATION WITH AN ATTORNEY IS ENCOURAGED WITH RESPECT TO ITS COMPLETION OR MODIFICATION

This document is intended to be used in conjunction with AIA Document B161, Owner-Architect Agreement for Designated Services

THIS DOCUMENT PROVIDES A COMPREHENSIVE SCHEDULE AND DESCRIPTIONS OF PROJECT PHASES AND SERVICES TO ENABLE THE OWNER AND THE ARCHITECT TO IDENTIFY THEIR RESPECTIVE RESPONSIBILITIES UNDER THE OWNER-ARCHITECT AGREEMENT. PARTICULAR SERVICES REQUIRED OR APPROPRIATE FOR ANY SPECIFIC PROJECT MUST BE IDENTIFIED BY COMPLETION OF THIS DOCUMENT, AND THE DESCRIPTIONS OF THOSE SERVICES MUST BE MODIFIED BY DELETIONS AND ADDITIONS WHERE APPROPRIATE.

Date:

In accordance with the Owner-Architect Agreement entered into as of

, the Owner and the Architect agree that the following designated services shall be provided by the parties as described herein, and that compensation for the services provided by the Architect shall be in accordance with the methods designated as applying thereto, and in the amounts provided in the Owner-Architect Agreement. The services specifically designated herein shall govern and shall supersede, to the extent of any conflict, all prior negotiations, representations or agreements, either written or oral.

Owner:

Architect:

Project:

The Owner and the Architect agree as set forth below.

SCHEDULE OF DESIGNATED SERVICES

PHASE 1: PREDESIGN SERVICES

PROJECT:

PROJECT #: **DATE:**

The Owner and the Architect, respectively, shall provide for this Phase of the Project those services appropriately designated and assigned below and described subsequently.

OWNER *(signed)* ____________________

ARCHITECT *(signed)* ____________________

	By Architect	By Architect, as Outside Services			By Owner, coordinated by Architect	By Owner			By Architect, as Additional Service*	Not to be provided	Method of Compensation
1.01 Project Administration											
1.02 Disciplines Coordination/Document Checking											
1.03 Agency Consulting/Review/Approval											
1.04 Owner-Supplied Data Coordination											
1.05 Programming											
1.06 Space Schematics/Flow Diagrams											
1.07 Existing Facilities Surveys											
1.08 Marketing Studies											
1.09 Economic Feasibility Studies											
1.10 Project Financing											
1.29 Project Development Scheduling											
1.31 Project Budgeting											
1.33 Presentations											

METHODS OF COMPENSATION

A = Multiple of Direct Salary Expense
B = Multiple of Direct Personnel Expense
C = Professional Fee Plus Expenses
D = Percentage of Construction Cost
E = Stipulated Sum
F = Hourly Billing Rates
G = Multiple of Amounts Billed to Architect
H = Other:

*Requires separate authorization and compensation

DESCRIPTION OF DESIGNATED SERVICES

PHASE 1—PREDESIGN SERVICES

In the Predesign Phase the Owner and Architect, respectively, shall provide those services designated in the Schedule of Designated Services necessary for the Architect to assist the Owner in establishing the Owner's program and the financial and time requirements and limitations for the Project prior to beginning design. The following descriptions shall apply to those services assigned in the Schedule of Designated Services as the responsibility of the party indicated therein.

1.01 Project Administration services consisting of predesign administrative functions including:

- **.01** Initial consultation in development of the Project
- **.02** Preparation of compensation estimates and professional services agreement(s)
- **.03** Project-related research
- **.04** Conferences
- **.05** Communications
- **.06** Travel time
- **.07** Progress reports
- **.08** Direction of the work of in-house architectural personnel.

1.02 Disciplines Coordination/Document Checking services consisting of predesign activities for:

- **.01** Coordination between the architectural work and the work of engineering and other involved disciplines for the Project.
- **.02** Review and checking of documents prepared for the Project.

1.03 Agency Consulting/Review/Approval services consisting of predesign activities including:

- **.01** Agency consultations
- **.02** Research of critical applicable regulations
- **.03** Research of community attitudes
- **.04** Preparation of written and graphic explanatory materials
- **.05** Appearances on Owner's behalf at agency and community meetings.

The services apply to applicable laws, statutes, regulations and codes of regulating entities and to reviews required of user or community groups with limited or no statutory authority but significant influence on approving agencies and individuals, including:

- **.06** Local political subdivisions
- **.07** Planning boards
- **.08** County agencies
- **.09** Regional agencies
- **.10** Federal agencies
- **.11** User organizations
- **.12** Community organizations
- **.13** Consumer interest organizations
- **.14** Environmental interest groups.

1.04 Owner-supplied Data Coordination services consisting of predesign activities including:

- **.01** Review and coordination of data furnished for the Project as a responsibility of the Owner
- **.02** Assistance to Owner in obtaining existing drawings and related data.

1.05 Programming services required to establish the following detailed requirements for the Project:

- **.01** Design objectives, limitations and criteria
- **.02** Space requirements
- **.03** Space relations
- **.04** Number and functional responsibilities of personnel
- **.05** Flexibility and expansibility
- **.06** Special equipment and systems
- **.07** Site requirements.

1.06 Space Schematics/Flow Diagrams services consisting of diagrammatic studies and pertinent descriptive text for:

- **.01** Internal functions
- **.02** Human, vehicular and material flow patterns
- **.03** General space allocations
- **.04** Analysis of operating functions
- **.05** Adjacency
- **.06** Special facilities and equipment
- **.07** Materials handling
- **.08** Flexibility and expansibility.

1.07 Existing Facilities Surveys services consisting of researching, assembling, reviewing and supplementing information for Projects involving alterations and additions to existing facilities or determining new space usage in conjunction with a new building program and including:

- **.01** Photography
- **.02** Field measurements
- **.03** Review of existing design data
- **.04** Analysis of existing structural capabilities
- **.05** Analysis of existing mechanical capabilities
- **.06** Analysis of existing electrical capabilities
- **.07** Review of existing drawings for critical inaccuracies, and the development of required measured drawings.

1.08 Marketing Studies services relating to determining the social, economic and political need and acceptability of the Project and consisting of:

- **.01** Determination with Owner of the scope, parameters, schedule and budget for marketing studies
- **.02** Identification, assembly, review and organization of existing pertinent data
- **.03** Arrangement of clearances for use of existing data
- **.04** Mail survey studies
- **.05** Personal survey studies
- **.06** Analysis of data
- **.07** Assistance in obtaining computerized analysis and modeling
- **.08** Computerized analysis and modeling
- **.09** Preparation of interim reports
- **.10** Preparation of final report
- **.11** Assistance in production of final report.

1.09 Economic Feasibility Studies services consisting of the preparation of economic analysis and feasibility evaluation of the Project based on estimates of:

.01 Total Project cost estimates
.02 Operating and owning cost estimates
.03 Financing requirements
.04 Cash flow for design, construction and operation
.05 Return on investment studies
.06 Equity requirements.

1.10 Project Financing services as required in connection with:

.01 Assistance to Owner in preparing and submitting data, supplementary drawings and documentation
.02 Research of financing availability
.03 Direct solicitation of financing sources by the Architect.

Project financing services are required for:

.04 Development costs
.05 Site control and/or acquisition
.06 Predesign and site analysis services
.07 Planning, design, documentation and bidding services
.08 Interim or construction financing
.09 Permanent or long-term financing.

1.29 Project Development Scheduling services during the Predesign Phase consisting of establishing a tentative schedule for predesign services, decision-making, design, documentation, contracting and construction, based on determination of the Architect's services, the Owner's responsibilities and proposed design and construction procedures.

1.31 Project Budgeting services relating to development of a Predesign Phase Statement of Probable Construction Cost based on programming and scheduling studies and consisting of:

.01 Conversion of programmed requirements to net area requirements
.02 Development of initial approximate gross facility areas
.03 Evaluation of construction market conditions
.04 Application of unit cost data to gross areas
.05 Estimates of related costs such as site development, landscaping, utilities, services, furniture, equipment and design services.

1.33 Presentations services consisting of presentations of Predesign Phase analyses and recommendations by the Architect to the following client representatives:

.01 Owner
.02 Building committee(s)
.03 Staff committee(s)
.04 User group(s)
.05 Board(s) of Directors
.06 Financing entity (entities)
.07 Owner's consultant(s).

NOTE: *Preparation of special presentation materials is a service to be identified, if relevant, under Agency Consulting/Review/Approval (1.03) or as a supplemental service identified in Phase 9 of the Schedule of Designated Services under Renderings (9.62) or in Phase 9 under Model Construction (9.63).*

SCHEDULE OF DESIGNATED SERVICES

PHASE 2: SITE ANALYSIS SERVICES

PROJECT:

PROJECT #: **DATE:**

The Owner and the Architect, respectively, shall provide for this Phase of the Project those services appropriately designated and assigned below and described subsequently.

OWNER *(signed)*________________________

ARCHITECT *(signed)*________________________

	By Architect	By Architect, as Outside Services			By Owner, coordinated by Architect	By Owner			By Architect, as Additional Service*	Not to be provided	Method of Compensation
2.01 Project Administration											
2.02 Disciplines Coordination/Document Checking											
2.03 Agency Consulting/Review/Approval											
2.04 Owner-Supplied Data Coordination											
2.13 Site Analysis and Selection											
2.14 Site Development Planning											
2.15 Detailed Site Utilization Studies											
2.16 On-Site Utility Studies											
2.17 Off-Site Utility Studies											
2.18 Environmental Studies and Reports											
2.19 Zoning Processing Assistance											
2.29 Project Development Scheduling											
2.31 Project Budgeting											
2.33 Presentations											

METHODS OF COMPENSATION

A = Multiple of Direct Salary Expense
B = Multiple of Direct Personnel Expense
C = Professional Fee Plus Expenses
D = Percentage of Construction Cost
E = Stipulated Sum
F = Hourly Billing Rates
G = Multiple of Amounts Billed to Architect
H = Other:

*Requires separate authorization and compensation

PHASE 2—SITE ANALYSIS SERVICES

In the Site Analysis Phase the Owner and the Architect, respectively, shall provide those services designated in the Schedule of Designated Services, necessary to establish site-related limitations and requirements for the Project. Some of the described services will be provided during predesign activities and some during other Phases of Project design. Grouping herein is for convenience to identify the site analysis services required for the Project. The following descriptions shall apply to those services assigned in the Schedule of Designated Services as the responsibility of the party indicated therein.

2.01 **Project Administration** services consisting of site-related administrative functions including:

- **.01** Initial site consultation
- **.02** Site-related research
- **.03** Conferences
- **.04** Communications
- **.05** Travel time
- **.06** Progress reports
- **.07** Direction of the work of in-house architectural personnel.

2.02 **Disciplines Coordination/Document Checking** services consisting of site analysis activities for:

- **.01** Coordination between the architectural work and the work of engineering and other involved disciplines for the Project
- **.02** Review and checking of documents prepared for the Project.

2.03 **Agency Consulting/Review/Approval** services consisting of site-related activities including:

- **.01** Agency consultations
- **.02** Research of critical applicable regulations
- **.03** Research of community attitudes
- **.04** Preparation of written and graphic explanatory materials
- **.05** Appearances on Owner's behalf at agency and community meetings.

The services apply to applicable laws, statutes, regulations and codes of regulating entities and to reviews required of user or community groups with limited or no statutory authority but significant influence on approving agencies and individuals, including:

- **.06** Local political subdivisions
- **.07** Planning boards
- **.08** County agencies
- **.09** Regional agencies
- **.10** Federal agencies
- **.11** User organizations
- **.12** Community organizations
- **.13** Consumer interest organizations
- **.14** Environmental interest groups.

2.04 Owner-supplied Data Coordination services consisting of site-related activities including:

- **.01** Review and coordination of data furnished for the Project as a responsibility of the Owner
- **.02** Site visits required to establish criteria
- **.03** Assistance to the Owner in obtaining existing title information and contracting for surveys, surface site data and subsurface site data
- **.04** Assistance to Owner in contracting for new boundary, surface and subsurface site data
- **.05** Site photography.

2.13 Site Analysis and Selection services consisting of:

- **.01** Identification of potential site(s)
- **.02** On-site investigations
- **.03** Movement systems, traffic and parking studies
- **.04** Topography analysis
- **.05** Analysis of deed, zoning and other legal restrictions
- **.06** Studies of availability of labor-force to staff Owner's facility
- **.07** Studies of availability of construction materials, equipment and labor
- **.08** Studies of construction market
- **.09** Overall site analysis and evaluation
- **.10** Comparative site studies.

2.14 Site Development Planning services consisting of preliminary site analysis, and preparation and comparative evaluation of conceptual site development designs, based on:

- **.01** Land utilization
- **.02** Structures placement
- **.03** Facilities development
- **.04** Development phasing
- **.05** Movement systems, circulation and parking
- **.06** Utilities systems
- **.07** Surface and subsurface conditions
- **.07** Ecological requirements
- **.09** Deeds, zoning and other legal restrictions
- **.10** Landscape concepts and forms.

2.15 Detailed Site Utilization Studies consisting of detailed site analysis, based on the approved conceptual site development design, including:

- **.01** Land utilization
- **.02** Structures placement
- **.03** Facilities development
- **.04** Development phasing
- **.05** Movement systems, circulation and parking
- **.06** Utilities systems
- **.07** Surface and subsurface conditions
- **.08** Review of soils report
- **.09** Vegetation
- **.10** Slope analysis
- **.11** Ecological studies
- **.12** Deeds, zoning and other legal restrictions
- **.13** Landscape forms and materials.

2.16 On-site Utility Studies services consisting of establishing requirements and preparing initial designs for on-site:

.01 Electrical service and distribution
.02 Gas service and distribution
.03 Water supply and distribution
.04 Site drainage
.05 Sanitary sewer collection and disposal
.06 Process waste water treatment
.07 Storm water collection and disposal
.08 Central-plant mechanical systems
.09 Fire systems
.10 Emergency systems
.11 Security
.12 Pollution control
.13 Site illumination
.14 Communications systems.

NOTE: *Master planning of on-site utilities for future development is a service to be identified, if relevant, in Phase 9 of the Schedule of Designated Services under Special Studies (9.61).*

2.17 Off-site Utility Studies services consisting of:

.01 Confirmation of location, size and adequacy of utilities serving the site
.02 Determination of requirements for connections to utilities
.03 Planning for off-site utility extensions and facilities
.04 Design of off-site utility extensions and facilities.

2.18 Environmental Studies and Reports services consisting of:

.01 Determination of need or requirements for environmental monitoring, assessment and/or impact statements
.02 Ecological studies
.03 Preparation of environmental assessment reports
.04 Preparation of environmental impact reports
.05 Attendance at public meetings and hearings
.06 Presentations to governing authorities.

NOTE: *Environmental Monitoring is a service to be identified, if relevant, in Phase 9 of the Schedule of Designated Services (9.69).*

2.19 Zoning Processing Assistance services consisting of:

.01 Assistance in preparing applications
.02 Development of supporting data
.03 Preparation of presentation materials
.04 Attendance at public meetings and hearings.

2.29 Project Development Scheduling services during the Site Analysis Phase consisting of establishing a tentative schedule for site analysis services, site-related decision-making, design, documentation, contracting and construction based on determination of the Architect's services, the Owner's responsibilities and proposed design and construction procedures.

2.31 Project Budgeting services relating to development of a Statement of Probable Construction Cost for site-related work based on programming and scheduling of site work and consisting of the application of unit cost data for:

.01 Site grading
.02 On-site utilities
.03 Utilities connections
.04 Off-site utilities work
.05 Drainage
.06 Paving
.07 Site lighting
.08 Lawn and landscaping work
.09 Site furniture.

2.33 Presentations services consisting of presentations of Site Analysis studies, reports and data by the Architect to the following client representatives:

.01 Owner
.02 Building committee(s)
.03 Staff committee(s)
.04 User group(s)
.05 Board(s) of Directors
.06 Financing entity(entities).

NOTE: *Preparation of special presentation materials is a service to be identified, if relevant, under Agency Consulting/Review/Approval (2.03) or as a supplemental service identified in Phase 9 of the Schedule of Designated Services under Renderings (9.62) or in Phase 9 under Model Construction (9.63).*

SCHEDULE OF DESIGNATED SERVICES — PHASE 3: SCHEMATIC DESIGN SERVICES

PROJECT:

PROJECT #: **DATE:**

The Owner and the Architect, respectively, shall provide for this Phase of the Project those services appropriately designated and assigned below and described subsequently.

OWNER *(signed)* ____________________________

ARCHITECT *(signed)* ____________________________

	By Architect	By Architect, as Outside Services			By Owner, coordinated by Architect	By Owner			By Architect, as Additional Service*	Not to be provided	Method of Compensation
3.01 Project Administration											
3.02 Disciplines Coordination/Document Checking											
3.03 Agency Consulting/Review/Approval											
3.04 Owner-Supplied Data Coordination											
3.21 Architectural Design/Documentation											
3.22 Structural Design/Documentation											
3.23 Mechanical Design/Documentation											
3.24 Electrical Design/Documentation											
3.25 Civil Design/Documentation											
3.26 Landscape Design/Documentation											
3.27 Interior Design/Documentation											
3.28 Materials Research/Specifications											
3.29 Project Development Scheduling											
3.32 Statement of Probable Construction Cost											
3.33 Presentations											

METHODS OF COMPENSATION

A = Multiple of Direct Salary Expense
B = Multiple of Direct Personnel Expense
C = Professional Fee Plus Expenses
D = Percentage of Construction Cost
E = Stipulated Sum
F = Hourly Billing Rates
G = Multiple of Amounts Billed to Architect
H = Other:

*Requires separate authorization and compensation

DESCRIPTION OF DESIGNATED SERVICES

PHASE 3—SCHEMATIC DESIGN SERVICES

In the Schematic Design Phase the Owner and the Architect, respectively, shall provide those services designated in the Schedule of Designated Services necessary to prepare Schematic Design Documents consisting of drawings and other documents illustrating the general scope, scale and relationship of Project components for approval by the Owner. Designs will be conceptual in character and based on the requirements developed under previous phases and approved by the Owner, or on program requirements provided by the Owner and reviewed and agreed upon by the Architect. The following descriptions shall apply to those services assigned in the Schedule of Designated Services as the responsibility of the party indicated therein.

3.01 Project Administration services consisting of schematic design administrative functions including:

.01 Consultation
.02 Research
.03 Conferences
.04 Communications
.05 Travel time
.06 Progress reports
.07 Direction of the work of in-house architectural personnel.

3.02 Disciplines Coordination/Document Checking services consisting of schematic design activities for:

.01 Coordination between the architectural work and the work of engineering and other involved disciplines for the Project
.02 Review and checking of documents prepared for the Project.

3.03 Agency Consulting/Review/Approval services consisting of schematic design activities including:

.01 Agency consultations
.02 Research of critical applicable regulations
.03 Preparation of written and graphic explanatory materials
.04 Appearances on Owner's behalf at agency and community meetings.

The services apply to applicable laws, statutes, regulations and codes of regulating entities and to reviews required of user or community groups with limited or no statutory authority but significant influence on approving agencies and individuals, including:

.05 Local political subdivisions
.06 Planning boards
.07 County agencies
.08 Regional agencies
.09 Federal agencies
.10 User organizations
.11 Community organizations
.12 Consumer interest organizations
.13 Environmental interest groups.

3.04 Owner-supplied Data Coordination services consisting of schematic design activities including:

.01 Review and coordination of data furnished for the Project as a responsibility of the Owner
.02 Assistance in establishing criteria
.03 Assistance in obtaining data.

3.21 **Architectural Design/Documentation** services during the Schematic Design Phase responding to program requirements and consisting of preparation of:

- **.01** Conceptual site and building plans
- **.02** Preliminary sections and elevations
- **.03** Preliminary selection of building systems and materials
- **.04** Development of approximate dimensions, areas and volumes
- **.05** Perspective sketch(es)
- **.06** Study model(s).

3.22 **Structural Design/Documentation** services during the Schematic Design Phase consisting of recommendations regarding basic structural materials and systems, analyses, and development of conceptual design solutions for:

- **.01** A predetermined structural system
- **.02** Alternate structural systems.

3.23 **Mechanical Design/Documentation** services during the Schematic Design Phase consisting of consideration of alternate materials, systems and equipment, and development of conceptual design solutions for:

- **.01** Energy source(s)
- **.02** Energy conservation
- **.03** Heating and ventilating
- **.04** Air conditioning
- **.05** Plumbing
- **.06** Fire protection
- **.07** Special mechanical systems
- **.08** Process systems
- **.09** General space requirements.

3.24 **Electrical Design/Documentation** services during the Schematic Design Phase consisting of consideration of alternate systems, recommendations regarding basic electrical materials, systems and equipment, analyses, and development of conceptual design solutions for:

- **.01** Power service and distribution
- **.02** Lighting
- **.03** Telephones
- **.04** Fire detection and alarms
- **.05** Security systems
- **.06** Electronic communications
- **.07** Special electrical systems
- **.08** General space requirements.

3.25 **Civil Design/Documentation** services during the Schematic Design Phase consisting of consideration of alternate materials and systems and development of conceptual design solutions for:

- **.01** On-site utility systems
- **.02** Off-site utilities work
- **.03** Fire protection systems
- **.04** Drainage systems
- **.05** Paving.

3.26 Landscape Design/Documentation services during the Schematic Design Phase consisting of consideration of alternate materials, systems and equipment and development of conceptual design solutions for land forms, lawns and plantings based on program requirements, physical site characteristics, design objectives and environmental determinants.

3.27 Interior Design/Documentation services during the Schematic Design Phase consisting of space allocation and utilization plans based on functional relationships, consideration of alternate materials, systems and equipment and development of conceptual design solutions for architectural, mechanical, electrical and equipment requirements in order to establish:

.01 Partition locations
.02 Furniture and equipment layouts.

3.28 Materials Research/Specifications services during the Schematic Design Phase consisting of:

.01 Identification of potential architectural materials, systems and equipment and their criteria and quality standards consistent with the conceptual design.
.02 Investigation of availability and suitability of alternative architectural materials, systems and equipment
.03 Coordination of similar activities of other disciplines.

3.29 Project Development Scheduling services during the Schematic Design Phase consisting of reviewing and updating previously established Project schedules or initial development of Project schedules (if not previously established) for decision-making, design, documentation, contracting and construction.

3.32 Statement of Probable Construction Cost services during the Schematic Design Phase consisting of development of a probable construction cost range for the Project based on the most recent schematic design studies, current and historic area, volume, or other unit costs, expected Project delivery process, and appropriate contingencies.

3.33 Presentations services consisting of presentations of Schematic Design Documents by the Architect to the following client representatives:

.01 Owner
.02 Building committee(s)
.03 Staff committee(s)
.04 User group(s)
.05 Board(s) of Directors
.06 Financing entity(enties)
.07 Owner's consultant(s).

NOTE: *Preparation of special presentation materials is a service to be identified, if relevant, under Agency Consulting/Review/Approval (3.03) or as a supplemental service identified in Phase 9 of the Schedule of Designated Services under Renderings (9.62) or in Phase 9 under Model Construction (9.63).*

SCHEDULE OF DESIGNATED SERVICES PHASE 4: DESIGN DEVELOPMENT SERVICES

PROJECT:

PROJECT #: **DATE:**

The Owner and the Architect, respectively, shall provide for this Phase of the Project those services appropriately designated and assigned below and described subsequently.

OWNER *(signed)*____________________

ARCHITECT *(signed)*____________________

	By Architect	By Architect, as Outside Services			By Owner, coordinated by Architect	By Owner			By Architect, as Additonal Service*	Not to be provided	Method of Compensation
4.01 Project Administration											
4.02 Disciplines Coordination/Document Checking											
4.03 Agency Consulting/Review/Approval											
4.04 Owner-Supplied Data Coordination											
4.21 Architectural Design/Documentation											
4.22 Structural Design/Documentation											
4.23 Mechanical Design/Documentation											
4.24 Electrical Design/Documentation											
4.25 Civil Design/Documentation											
4.26 Landscape Design/Documentation											
4.27 Interior Design/Documentation											
4.28 Materials Research/Specifications											
4.29 Project Development Scheduling											
4.32 Statement of Probable Construction Cost											
4.33 Presentations											

METHODS OF COMPENSATION

A = Multiple of Direct Salary Expense
B = Multiple of Direct Personnel Expense
C = Professional Fee Plus Expenses
D = Percentage of Construction Cost
E = Stipulated Sum
F = Hourly Billing Rates
G = Multiple of Amounts Billed to Architect
H = Other:

*Requires separate authorization and compensation

DESCRIPTION OF DESIGNATED SERVICES

PHASE 4—DESIGN DEVELOPMENT SERVICES

In the Design Development Phase the Owner and the Architect, respectively, shall provide those services designated in the Schedule of Designated Services necessary to prepare from the approved Schematic Design Documents, for approval by the Owner, the Design Development Documents consisting of drawings and other documents to fix and describe the size and character of the entire Project, including architectural, structural, mechanical and electrical systems, materials and such other elements as may be appropriate. Consideration shall be given to availability of materials, equipment and labor, construction sequencing and scheduling, economic analysis of construction and operations, user safety and maintenance requirements, and energy conservation. The following descriptions shall apply to those services assigned in the Schedule of Designated Services as the responsibility of the party indicated therein.

4.01 Project Administration services consisting of design development administrative functions including:

- **.01** Consultation
- **.02** Research
- **.03** Conferences
- **.04** Communications
- **.05** Travel time
- **.06** Progress reports
- **.07** Direction of the work of in-house architectural personnel.

4.02 Disciplines Coordination/Document Checking services consisting of design development activities for:

- **.01** Coordination between the architectural work and the work of engineering and other involved disciplines for the Project
- **.02** Review and checking of documents prepared for the Project.

4.03 Agency Consulting/Review/Approval services consisting of design development activities including:

- **.01** Agency consultations
- **.02** Research of applicable regulations
- **.03** Preparation of written and graphic explanatory materials
- **.04** Appearances on Owner's behalf at agency and community meetings.

The services apply to applicable laws, statutes, regulations and codes of regulating entities and to reviews required of user or community groups with limited or no statutory authority but significant on approving agencies and individuals, including:

- **.05** Local political subdivisions
- **.06** Planning boards
- **.07** County agencies
- **.08** Regional agencies
- **.09** Federal agencies
- **.10** User organizations
- **.11** Community organizations
- **.12** Consumer interest organizations
- **.13** Environmental interest groups.

4.04 Owner-supplied Data Coordination services consisting of design development activities including:

- **.01** Review and coordination of data furnished for the Project as a responsibility of the Owner
- **.02** Assistance in establishing criteria
- **.03** Assistance in obtaining data.

4.21 Architectural Design/Documentation services during the Design Development Phase consisting of continued development and expansion of architectural Schematic Design Documents to establish the final scope, relationships, forms, size and appearance of the Project through:

- **.01** Plans, sections and elevations
- **.02** Typical construction details
- **.03** Three dimensional sketch(es)
- **.04** Study model(s)
- **.05** Final materials selections
- **.06** Equipment layouts

NOTE: *Furniture, furnishings and interior equipment layouts are a service to be identified, if relevant, under Interior Design/Documentation (4.27).*

4.22 Structural Design/Documentation services during the Design Development Phase consisting of continued development of the specific structural system(s) and Schematic Design Documents in sufficient detail to establish:

- **.01** Basic structural system and dimensions
- **.02** Final structural design criteria
- **.03** Foundation design criteria
- **.04** Preliminary sizing of major structural components
- **.05** Critical coordination clearances
- **.06** Outline Specifications or materials lists.

4.23 Mechanical Design/Documentation services during the Design Development Phase consisting of continued development and expansion of mechanical Schematic Design Documents and development of outline Specifications or materials lists to establish:

- **.01** Approximate equipment sizes and capacities
- **.02** Preliminary equipment layouts
- **.03** Required space for equipment
- **.04** Required chases and clearances
- **.05** Acoustical and vibration control
- **.06** Visual impacts
- **.07** Energy conservation measures.

4.24 Electrical Design/Documentation services during the Design Development Phase consisting of continued development and expansion of electrical Schematic Design Documents and development of outline Specifications or materials lists to establish:

- **.01** Criteria for lighting, electrical and communications systems
- **.02** Approximate sizes and capacities of major components
- **.03** Preliminary equipment layouts
- **.04** Required space for equipment
- **.05** Required chases and clearances.

4.25 **Civil Design/Documentation** services during the Design Development Phase consisting of continued development and expansion of civil Schematic Design Documents and development of outline Specifications or materials lists to establish the final scope and preliminary details for on-site and off-site civil engineering work.

4.26 **Landscape Design/Documentation** services during the Design Development Phase consisting of continued development and expansion of landscape Schematic Design Documents and development of outline Specifications or materials lists to establish final scope and preliminary details for landscape work.

4.27 **Interior Design/Documentation** services during the Design Development Phase consisting of continued development and expansion of interior Schematic Design Documents and development of outline Specifications or materials lists to establish final scope and preliminary details relative to:

.01 Interior construction of the Project
.02 Special interior design features
.03 Furniture, furnishings and equipment selections
.04 Materials and finishes and colors.

4.28 **Materials Research/Specifications** services during the Design Development Phase consisting of activities by in-house architectural personnel in:

.01 Presentation for Owner's approval of proposed General and Supplementary Conditions of the Contract for Construction for Owner's approval
.02 Development of architectural outline Specifications or itemized lists and brief form identification of significant architectural materials, systems and equipment and their criteria and quality standards
.03 Coordination of similar activities of other disciplines
.04 Production of design manual including design criteria and outline Specifications or materials lists.

4.29 **Project Development Scheduling** services during the Design Development Phase consisting of reviewing and updating previously established schedules for the Project.

4.32 **Statement of Probable Construction Cost** services during the Design Development Phase consisting of updating and refining the Schematic Design Phase Statement of Probable Construction Cost of the Project, taking into consideration:

.01 Availability of materials and labor
.02 Project delivery procedures
.03 Construction sequencing and scheduling
.04 Changes in scope of the Project
.05 Adjustments in quality standards.

4.33 **Presentations** services consisting of presentations of Design Development Drawings and other documents by the Architect to the following client representatives:

.01 Owner
.02 Building committee(s)
.03 Staff committee(s)
.04 User group(s)
.05 Board(s) of Directors
.06 Financing entity(entities)
.07 Owner's consultant(s).

NOTE: *Preparation of special presentation materials is a service to be identified, if relevant, under Agency Consulting/Review/Approval (4.03) or as a supplemental service identified in Phase 9 of the Schedule of Designated Services under Renderings (9.62) or in Phase 9 under Model Construction (9.63).*

SCHEDULE OF DESIGNATED SERVICES

PHASE 5: CONSTRUCTION DOCUMENTS SERVICES

PROJECT:

PROJECT #: **DATE:**

The Owner and the Architect, respectively, shall provide for this Phase of the Project those services appropriately designated and assigned below and described subsequently.

OWNER *(signed)* ____________________

ARCHITECT *(signed)* ____________________

	By Architect	By Architect, as Outside Services			By Owner, coordinated by Architect	By Owner			By Architect, as Additional Service*	Not to be provided	Method of Compensation
5.01 Project Administration											
5.02 Disciplines Coordination/Document Checking											
5.03 Agency Consulting/Review/Approval											
5.04 Owner-Supplied Data Coordination											
5.21 Architectural Design/Documentation											
5.22 Structural Design/Documentation											
5.23 Mechanical Design/Documentation											
5.24 Electrical Design/Documentation											
5.25 Civil Design/Documentation											
5.26 Landscape Design/Documentation											
5.27 Interior Design/Documentation											
5.28 Materials Research/Specifications											
5.30 Special Bidding Documents/Scheduling											
5.32 Statement of Probable Construction Cost											
5.33 Presentations											

METHODS OF COMPENSATION

A = Multiple of Direct Salary Expense
B = Multiple of Direct Personnel Expense
C = Professional Fee Plus Expenses
D = Percentage of Construction Cost
E = Stipulated Sum
F = Hourly Billing Rates
G = Multiple of Amounts Billed to Architect
H = Other:

*Requires separate authorization and compensation

DESCRIPTION OF DESIGNATED SERVICES

PHASE 5—CONSTRUCTION DOCUMENTS SERVICES

In the Construction Documents Phase the Owner and the Architect, respectively, shall provide those services designated in the Schedule of Designated Services necessary to prepare, from the approved Design Development Documents, for approval by the Owner, Construction Documents consisting of Drawings, Specifications and other documents setting forth in detail the requirements for construction of the Project and bidding and contracting for the construction of the Project. The following descriptions shall apply to those services assigned in the Schedule of Designated Services as the responsibility of the party indicated therein.

5.01 Project Administration services consisting of construction documents administrative functions including:

- **.01** Consultation
- **.02** Research
- **.03** Conferences
- **.04** Communications
- **.05** Travel time
- **.06** Progress reports
- **.07** Direction of the work in-house architectural personnel.

5.02 Disciplines Coordination/Document Checking services consisting of construction documents activities for:

- **.01** Coordination between the architectural work and the work of engineering and other involved disciplines for the Project
- **.02** Review and checking of documents prepared for the Project.

5.03 Agency Consulting/Review/Approval services relating to applicable laws, statutes, regulations and codes of regulating entities and consisting of Construction Documents Phase activities for:

- **.01** Agency consultations
- **.02** Research of applicable regulations
- **.03** Appearances on Owner's behalf at agency meetings.

5.04 Owner-supplied Data Coordination services consisting of activities relating to the construction documents including:

- **.01** Review and coordination of data furnished for the Project as a responsibility of the Owner
- **.02** Assistance in establishing criteria
- **.03** Assistance in obtaining data.

5.21 Architectural Design/Documentation services during the Construction Documents Phase consisting of preparation of Drawings based on approved Design Development Documents setting forth in detail the architectural construction requirements for the Project.

NOTE: *Preparation of architectural Specifications is a service, to be identified, if relevant, under Materials Research/Specifications (5.28).*

5.22 **Structural Design/Documentation** services during the Construction Documents Phase consisting of preparation of final structural engineering calculations, Drawings and Specifications based on approved Design Development Documents, setting forth in detail the structural construction requirements for the Project.

5.23 **Mechanical Design/Documentation** services during the Construction Documents Phase consisting of preparation of final mechanical engineering calculations, Drawings and Specifications based on approved Design Development Documents, setting forth in detail the mechanical construction requirements for the Project.

5.24 **Electrical Design/Documentation** services during the Construction Documents Phase consisting of preparation of final electrical engineering calculations, Drawings and Specifications based on approved Design Development Documents, setting forth in detail the electrical construction requirements for the Project.

5.25 **Civil Design/Documentation** services during the Construction Documents Phase consisting of preparation of final civil engineering calculations, Drawings and Specifications based on approved Design Development Documents, setting forth in detail the civil construction requirements for the Project.

5.26 **Landscape Design/Documentation** services during the Construction Documents Phase consisting of preparation of Drawings and Specifications based on approved Design Development Documents, setting forth in detail the landscape requirements for the Project.

5.27 **Interior Design/Documentation** services during the Construction Documents Phase consisting of preparation of Drawings, Specifications and furnishings, furniture and equipment schedules based on approved Design Development Documents, setting forth in detail the requirements for interior construction and for furniture, furnishings and equipment for the Project.

5.28 **Materials Research/Specifications** during the Construction Documents Phase consisting of activities of in-house architectural personnel in:

- **.01** Assistance to the Owner in development and preparation of Bidding Documents which describe the time, place and conditions of bidding, bidding forms, and the form(s) of Agreement between the Owner and Contractor(s)
- **.02** Assistance to the Owner in development and preparation of the Conditions of the Contract (General, Supplementary, and other Conditions)
- **.03** Development and preparation of architectural Specifications describing materials, systems and equipment, workmanship, quality and performance criteria required for the construction of the Project
- **.04** Coordination of the development of Specifications by other disciplines
- **.05** Compilation of Project Manual including Conditions of the Contract, Bidding Documents and Specifications.

5.30 Special Bidding Documents/Scheduling services during the Construction Documents Phase consisting of:

- **.01** Preparation and coordination of special Drawings and Specifications for obtaining bids or prices on alternate subdivisions of the Work
- **.02** Preparation and coordination of special Drawings and Specifications for obtaining alternate bids or prices on changes in the scope of the Work
- **.03** Preparation and coordination of Drawings, Specifications, Bidding Documents and schedules for out-of-sequence bidding or pricing of subdivisions of the Work
- **.04** Preparation and coordination of Drawings, Specifications and Bidding Documents for multiple prime contracts for subdivisions of the Work
- **.05** Review and update of established schedules for the Project.

5.32 Statement of Probable Construction Cost services during the Construction Documents Phase consisting of advising the Owner of any adjustments to, and, when the Construction Documents are approximately 90 percent complete, updating of the Design Development Phase Statement of Probable Construction Cost of the Project, taking into account:

- **.01** Changes in materials, systems or details of construction which have occurred during preparation of the Construction Documents
- **.02** Known changes in the cost of materials, labor and services since preparation of the previous Statement of Probable Construction Cost
- **.03** Adjustments for known or anticipated changes in the bidding market relative to the Project.

5.33 Presentations services consisting of presentations of Construction Documents and special presentation graphic displays by the Architect to the following client representatives:

- **.01** Owner
- **.02** Building committee(s)
- **.03** Staff committee(s)
- **.04** User group(s)
- **.05** Board(s) of Directors
- **.06** Financing entity(entities)
- **.07** Owner's consultant(s).

NOTE: *Preparation of special presentation materials is a service to be identified, if relevant, under Agency Consulting/Review/Approval (5.03) or as a supplemental service identified in Phase 9 of the Schedule of Designated Services under Renderings (9.62) or in Phase 9 under Model Construction (9.63).*

SCHEDULE OF DESIGNATED SERVICES

PHASE 6: BIDDING OR NEGOTIATIONS SERVICES

PROJECT:

PROJECT #: **DATE:**

The Owner and the Architect, respectively, shall provide for this Phase of the Project those services appropriately designated and assigned below and described subsequently.

OWNER *(signed)* ____________________

ARCHITECT *(signed)* ____________________

	By Architect	By Architect, as Outside Services			By Owner, coordinated by Architect	By Owner			By Architect, as Additional Service*	Not to be provided	Method of Compensation
6.01 Project Administration											
6.02 Disciplines Coordination/Document Checking											
6.03 Agency Consulting/Review/Approval											
6.04 Owner-Supplied Data Coordination											
6.34 Bidding Materials											
6.35 Addenda											
6.36 Bidding/Negotiations											
6.37 Analysis of Alternates/Substitutions											
6.38 Special Bidding Services											
6.39 Bid Evaluation											
6.40 Construction Contract Agreements											

METHODS OF COMPENSATION

A = Multiple of Direct Salary Expense
B = Multiple of Direct Personnel Expense
C = Professional Fee Plus Expenses
D = Percentage of Construction Cost
E = Stipulated Sum
F = Hourly Billing Rates
G = Multiple of Amounts Billed to Architect
H = Other:

*Requires separate authorization and compensation

DESCRIPTION OF DESIGNATED SERVICES

PHASE 6—BIDDING OR NEGOTIATIONS SERVICES

In the Bidding or Negotiations Phase the Owner and the Architect, respectively, following the Owner's approval of the Construction Documents and of the most recent Statement of Probable Construction Cost, shall provide those services designated in the Schedule of Designated Services necessary for the Architect to assist the Owner in obtaining bids or negotiated proposals and in awarding and preparing contracts for construction. In the case of phased construction, the Owner may authorize bidding and/or negotiation of portions of the Work prior to completion of the Construction Documents Phase. The following descriptions shall apply to those services assigned in the Schedule of Designated Services as the responsibility of the party indicated therein.

6.01 Project Administration services consisting of bidding or negotiation administrative functions including:

- **.01** Consultation
- **.02** Research
- **.03** Conferences
- **.04** Communications
- **.05** Travel time
- **.06** Progress reports
- **.07** Direction of the work of in-house architectural personnel.

6.02 Disciplines Coordination/Document Checking services consisting of bidding or negotiation activites for:

- **.01** Coordination between the architectural work and the work of engineering and other involved disciplines for the Project
- **.02** Review and checking of documents prepared for the Project.

6.03 Agency Consulting/Review/Approval services during the Bidding or Negotiations Phase relating to applicable laws, statutes, regulations and codes of regulating entities relating to the Owner's interests before construction begins.

6.04 Owner-supplied Data Coordination services consisting of activities relating to bidding or negotiation including:

- **.01** Review and coordination of data furnished for the Project as a responsibility of the Owner
- **.02** Assistance in establishing criteria
- **.03** Assistance in obtaining data

6.34 Bidding Materials services consisting of organizing and handling Bidding Documents for:

- **.01** Coordination
- **.02** Reproduction
- **.03** Completeness review
- **.04** Distribution
- **.05** Distribution records
- **.06** Retrieval
- **.07** Receipt and return of document deposits
- **.08** Review, repair and reassembly of returned materials.

6.35 **Addenda** services consisting of preparation and distribution of Addenda as may be required during bidding or negotiation and including supplementary Drawings, Specifications, instructions and notice(s) of changes in the bidding schedule and procedure.

6.36 **Bidding/Negotiations** services consisting of:

.01 Assistance to Owner in establishing list of Bidders or proposers
.02 Prequalification of Bidders or proposers
.03 Participation in pre-bid conferences
.04 Responses to questions from Bidders or proposers and clarifications or interpretations of the Bidding Documents
.05 Attendance at bid opening(s)
.06 Documentation and distribution of bidding results.

6.37 **Analysis of Alternates/Substitutions** services consisting of consideration, analyses, comparisons, and recommendations relative to alternates or substitutions proposed by Bidders or proposers either prior or subsequent to receipt of Bids or proposals.

6.38 **Special Bidding Services** consisting of:

.01 Attendance at bid openings, participation in negotiations, and documentation of decisions for multiple contracts or phased construction
.02 Technical evaluation of proposals for building systems
.03 Participation in detailed evaluation procedures for building systems proposals.

6.39 **Bid Evaluation** services consisting of:

.01 Validation of Bids or proposals
.02 Participation in reviews of Bids or proposals
.03 Evaluation of Bids or proposals
.04 Recommendation on award of Contract(s)
.05 Participation in negotiations prior to or following decisions on award of the Contract(s) for Construction.

6.40 **Construction Contract Agreements** services consisting of:

.01 Notification of Contract award(s)
.02 Assistance in preparation of construction contract Agreement forms for approval by Owner
.03 Preparation and distribution of sets of Contract Documents for execution by parties to the Contract(s)
.04 Receipt, distribution and processing, for Owner's approval, of required certificates of insurance, bonds and similar documents
.05 Preparation and distribution to Contractor(s), on behalf of the Owner, of notice(s) to proceed with the Work.

SCHEDULE OF DESIGNATED SERVICES

PHASE 7: CONSTRUCTION CONTRACT ADMINISTRATION SERVICES

PROJECT:

PROJECT #: **DATE:**

The Owner and the Architect, respectively, shall provide for this Phase of the Project those services appropriately designated and assigned below and described subsequently.

OWNER *(signed)* ____________________

ARCHITECT *(signed)* ____________________

	By Architect	By Architect, as Outside Services			By Owner, coordinated by Architect	By Owner			By Architect, as Additional Service*	Not to be provided	Method of Compensation
7.01 Project Administration											
7.02 Disciplines Coordination/Document Checking											
7.03 Agency Consulting/Review/Approval											
7.04 Owner-Supplied Data Coordination											
7.41 Office Construction Administration											
7.42 Construction Field Observation											
7.43 Project Representation											
7.44 Inspection Coordination											
7.45 Supplemental Documents											
7.46 Quotation Requests/Change Orders											
7.47 Project Schedule Monitoring											
7.48 Construction Cost Accounting											
7.49 Project Closeout											

METHODS OF COMPENSATION

A = Multiple of Direct Salary Expense
B = Multiple of Direct Personnel Expense
C = Professional Fee Plus Expenses
D = Percentage of Construction Cost
E = Stipulated Sum
F = Hourly Billing Rates
G = Multiple of Amounts Billed to Architect
H = Other:

*Requires separate authorization and compensation

DESCRIPTION OF DESIGNATED SERVICES

PHASE 7—CONSTRUCTION CONTRACT ADMINISTRATION SERVICES

In the Construction Contract Administration Phase the Owner and the Architect, respectively, shall provide those services designated in the Schedule of Designated Services necessary for the administration of the construction contract as set forth in the General Conditions of the Contract for Construction. Unless otherwise provided in the Schedule of Designated Services, the Architect's duties and responsibilities during construction shall be as set forth in the Agreement between Owner and Architect for Designated Services. The following descriptions shall apply to those services assigned in the Schedule of Designated Services as the responsibility of the party indicated therein.

7.01 Project Administration services consisting of construction contract administrative functions including:

- **.01** Consultation
- **.02** Research
- **.03** Conferences
- **.04** Communications
- **.05** Travel time
- **.06** Progress reports
- **.07** Coordination of out-of-normal-sequence construction operations
- **.08** Coordination of multiple-prime construction contracts
- **.09** Coordination of work by the Owner's forces
- **.10** Direction of the work of in-house architectural personnel.

7.02 Disciplines Coordination/Document Checking services consisting of construction contract administration activities for:

- **.01** Coordination between the architectural work and the work of engineering and other involved disciplines for the Project
- **.02** Review and checking of documents prepared for the Project.

7.03 Agency Consulting/Review/Approval services during the Construction Contract Administration Phase relating to applicable laws, statutes, regulations and codes of regulating entities relating to the Owner's interests during construction of the Project.

7.04 Owner-supplied Data Coordination services consisting of activities relating to construction contract administration including:

- **.01** Review and coordination of data furnished for the Project as a responsibility of the Owner
- **.02** Assistance in establishing criteria
- **.03** Assistance in obtaining data
- **.04** Coordination of delivery and installation for Owner-supplied equipment
- **.05** Coordination of delivery and installation of Owner-supplied furniture and furnishings.

7.41 Office Construction Administration services consisting of:

- **.01** Processing of submittals, including receipt, review of, and appropriate action on Shop Drawings, Product Data, Samples and other submittals required by the Contract Documents
- **.02** Distribution of submittals to Owner, Contractor and/or Architect's field representative as required
- **.03** Maintenance of master file of submittals
- **.04** Related communications.

7.42 Construction Field Observation services consisting of visits to the site at intervals appropriate to the stage of construction or as otherwise agreed in writing to become generally familiar with the progress and quality of the Work and to determine in general if the Work is proceeding in accordance with the Contract Documents, and preparing related reports and communications.

7.43 Project Representation services consisting of selection, employment and direction of:

- **.01** Full-time Project Representative(s)
- **.02** Part-time Project Representative(s).

7.44 Inspection Coordination services relating to independent inspection and testing agencies, consisting of:

- **.01** Administration and coordination of field testing required by the Contract Documents
- **.02** Recommending scope, standards, procedures and frequency of testing and inspections
- **.03** Arranging for testing and inspection on Owner's behalf
- **.04** Notifying inspection and testing agencies of status Work requiring testing and inspection
- **.05** Evaluating compliance by testing and inspection agencies with required scope, standards, procedures and frequency
- **.06** Review of reports on inspections and tests and notifications to Owner and Contractor(s) of observed deficiencies in the Work.

7.45 Supplemental Documents services consisting of:

- **.01** Preparation, reproduction and distribution of supplemental Drawings, Specifications and interpretations in response to requests for clarification by Contractor(s) or the Owner and as required by construction exigencies
- **.02** Forwarding Owner's instructions and providing guidance to the Contractor(s) on the Owner's behalf relative to changed requirements and schedule revisions.

7.46 Quotation Requests/Change Orders services consisting of:

- **.01** Preparation, reproduction and distribution of Drawings and Specifications to describe Work to be added, deleted or modified
- **.02** Review of proposals from Contractor(s) for reasonableness of quantities and costs of labor and materials
- **.02** Review and recommendations relative to changes in time for Substantial Completion
- **.04** Negotiations with Contractor(s) on Owner's behalf relative to costs of Work proposed to be added, deleted or modified
- **.05** Assisting in the preparation of appropriate Modifications of the Contract(s) for Construction
- **.06** Coordination of communications, approvals, notifications and record-keeping relative to changes in the Work.

7.47 Project Schedule Monitoring services consisting of monitoring the progress of the Contractor(s) relative to established schedules and making status reports to Owner.

7.48 Construction Cost Accounting services consisting of:

- **.01** Maintenance of records of payments on account of the Contract Sum and all changes thereto
- **.02** Evaluation of Applications for Payment and certification thereof
- **.03** Review and evaluation of expense data submitted by the Contractor(s) for Work performed under cost-plus-fee arrangements.

7.49 Project Closeout services initiated upon notice from the Contractor(s) that the Work, or a designated portion thereof which is acceptable to the Owner, is sufficiently complete, in accordance with the Contract Documents, to permit occupancy or utilization for the use for which it is intended, and consisting of:

- **.01** A detailed inspection with the Owner's representative for conformity of the Work to the Contract Documents to verify the list submitted by the Contractor(s) of items to be completed or corrected
- **.02** Determination of the amounts to be withheld until final completion
- **.03** Securing and receipt of consent of surety or sureties, if any, to reduction in or partial release of retainage or the making of final payment(s)
- **.04** Issuance of Certificate(s) of Substantial Completion
- **.05** Inspection(s) upon notice by the Contractor(s) that the Work is ready for final inspection and acceptance
- **.06** Notification to Owner and Contractor(s) of deficiencies found in follow-up inspection(s), if any
- **.07** Final inspection with the Owner's representative to verify final completion of the Work
- **.08** Receipt and transmittal of warranties, affidavits, receipts, releases and waivers of lien or bonds indemnifying the Owner against liens
- **.09** Securing and receipt of consent of surety or sureties, if any, to the making of final payment(s)
- **.10** Issuance of final Certificate(s) for Payment.

SCHEDULE OF DESIGNATED SERVICES	PHASE 8: POSTCONSTRUCTION SERVICES									
PROJECT: **PROJECT #:** **DATE:** The Owner and the Architect, respectively, shall provide for this Phase of the Project those services appropriately designated and assigned below and described subsequently. OWNER *(signed)* ______ ARCHITECT *(signed)* ______	By Architect	By Architect, as Outside Services			By Owner, coordinated by Architect	By Owner			By Architect, as Additional Service*	Not to be provided	Method of Compensation
8.01 Project Administration											
8.02 Disciplines Coordination/Document Checking											
8.03 Agency Consulting/Review/Approval											
8.04 Owner-Supplied Data Coordination											
8.50 Maintenance and Operational Programming											
8.51 Start-Up Assistance											
8.52 Record Drawings											
8.53 Warranty Review											
8.54 Postconstruction Evaluation											

METHODS OF COMPENSATION

A = Multiple of Direct Salary Expense
B = Multiple of Direct Personnel Expense
C = Professional Fee Plus Expenses
D = Percentage of Construction Cost
E = Stipulated Sum
F = Hourly Billing Rates
G = Multiple of Amounts Billed to Architect
H = Other:

*Requires separate authorization and compensation

DESCRIPTION OF DESIGNATED SERVICES

PHASE 8—POSTCONSTRUCTION SERVICES

In the Postconstruction Phase the Owner and the Architect, respectively, shall provide those services designated in the Schedule of Designated Services necessary to assist the Owner in the use and occupancy of the facility. The following descriptions shall apply to those services assigned in the Schedule of Designated Services as the responsibility of the party indicated therein.

8.01 **Project Administration** services consisting of postconstruction administrative functions including:

- **.01** Consultation
- **.02** Research
- **.03** Conferences
- **.04** Communications
- **.05** Travel time
- **.06** Progress reports
- **.07** Direction of the work of in-house architectural personnel.

8.02 **Disciplines Coordination/Document Checking** services consisting of postconstruction activities for:

- **.01** Coordination between the architectural work and the work of engineering and other involved disciplines for the Project
- **.02** Review and checking of documents prepared for use or occupancy of the Project.

8.03 **Agency Consulting/Review/Approval** services relating to applicable laws, statutes, regulations and codes of regulating entities that require compliance during postconstruction use, occupancy and operation.

8.04 **Owner-supplied Data Coordination** services consisting of postconstruction activities during occupancy and operation relative to Owner-supplied furniture, furnishings and equipment.

8.50 **Maintenance and Operational Programming** services consisting of:

- **.01** Assistance in the establishment by the Owner of in-house or contract program(s) of operation and maintenance of the physical plant and equipment
- **.02** Arranging for and coordinating instructions on operations and maintenance on equipment in conjunction with manufacturers' representatives
- **.03** Assistance in the preparation of operations and maintenance manual(s) for the Owner's use.

8.51 **Start-up Assistance** services consisting of:

- **.01** On-site observation, troubleshooting and assistance in the operation of building systems during initial occupancy
- **.02** Assistance in the training of the Owner's operation and maintenance personnel in proper operations, schedules and procedures
- **.03** Administration and coordination of remedial work by the Contractor(s) after final completion.

8.52 Record Drawings services consisting of:

- **.01** Making arrangements for obtaining from Contractor(s) and other parties information certified by them on all changes made during construction from the initial Contract Documents and on the location of concealed systems as installed during construction
- **.02** Review of general accuracy of information submitted and certified to by the Contractor(s)
- **.03** Preparation of record drawings, based on information furnished by the Contractor(s), including significant changes in the Work made during construction
- **.04** Transmittal of record drawings and general data, appropriately identified, to the Owner and others as directed.

8.53 Warranty Review services consisting of:

- **.01** Consultation and recommendation to the Owner during the duration of warranties in connection with inadequate performance of materials, systems and equipment under warranty
- **.02** Inspection(s) prior to expiration of the warranty period(s) to ascertain adequacy of performance of materials, systems and equipment
- **.03** Documenting defects or deficiencies and assisting the Owner in preparing instruction to the Contractor(s) for correction of noted defects.

8.54 Postconstruction Evaluation services consisting of a Project inspection at least one year after completion of construction; review with appropriate supervisory, operating and maintenance personnel; and analysis of operating costs and related data for evaluation of:

- **.01** The initial Project programming versus actual facility use
- **.02** The functional effectiveness of planned spaces and relationships
- **.03** The operational effectiveness of systems and materials installed
- **.04** Efficiency of the design and construction delivery processes used.

SCHEDULE OF DESIGNATED SERVICES

PHASE 9: SUPPLEMENTAL SERVICES

PROJECT:

PROJECT #: **DATE:**

The Owner and the Architect, respectively, shall provide for this Phase of the Project those services appropriately designated and assigned below and described subsequently.

OWNER *(signed)* ______________________________

ARCHITECT *(signed)* ______________________________

	By Architect	By Architect, as Outside Services			By Owner, coordinated by Architect	By Owner			By Architect, as Additional Service*	Not to be provided	Method of Compensation
9.61 Special Studies											
9.62 Renderings											
9.63 Model Construction											
9.64 Life Cycle Cost Analysis											
9.65 Value Analysis											
9.66 Quantity Surveys											
9.67 Detailed Construction Cost Estimates											
9.68 Energy Studies											
9.69 Environmental Monitoring											
9.70 Tenant-Related Services											
9.71 Graphics Design											
9.72 Fine Arts and Crafts Services											
9.73 Special Furnishings Design											
9.74 Non-Building Equipment Selections											
9.75 Project Promotion/Public Relations											
9.76 Leasing Brochures											
9.77 Expert Witness											
9.78 Computer Applications											

METHODS OF COMPENSATION

A = Multiple of Direct Salary Expense
B = Multiple of Direct Personnel Expense
C = Professional Fee Plus Expenses
D = Percentage of Construction Cost
E = Stipulated Sum
F = Hourly Billing Rates
G = Multiple of Amounts Billed to Architect
H = Other:

*Requires separate authorization and compensation

SCHEDULE OF DESIGNATED SERVICES

PHASE 9: SUPPLEMENTAL SERVICES (continued)

PROJECT:

PROJECT #: **DATE:**

The Owner and the Architect, respectively, shall provide for this Phase of the Project those services appropriately designated and assigned below and described subsequently.

OWNER *(signed)*____________________________

ARCHITECT *(signed)*____________________________

	By Architect	By Architect, as Outside Services			By Owner, coordinated by Architect	By Owner			By Architect, as Additional Service*	Not to be provided	Method of Compensation
9.79 Materials and Systems Testing											
9.80 Demolition Services											
9.81 Mock-Up Services											
9.82 Still Photography											
9.83 Motion Pictures and Videotape											
9.84 Coordination with Non-Design Professionals											
9.85 Special Disciplines Consultation											
9.86 Special Building Type Consultation											

METHODS OF COMPENSATION

A = Multiple of Direct Salary Expense
B = Multiple of Direct Personnel Expense
C = Professional Fee Plus Expenses
D = Percentage of Construction Cost
E = Stipulated Sum
F = Hourly Billing Rates
G = Multiple of Amounts Billed to Architect
H = Other:

*Requires separate authorization and compensation

DESCRIPTION OF DESIGNATED SERVICES

PHASE 9—SUPPLEMENTAL SERVICES

In addition to the generally sequential services chronologically arranged and described in Phases 1 through 8, the Owner and Architect, respectively, shall provide supplemental services designated in the Schedule of Designated Services. These nonsequential services may be provided during a single Phase of services or during several phases. The following descriptions shall apply to those services assigned in the Schedule of Designated Services as the responsibility of the party indicated therein.

9.61 **Special Studies** services consisting of investigation, research and analysis of the Owner's special requirements for the Project and documentation of findings, conclusions and recommendations for:

.01 Master planning to provide design services relative to future facilities, systems and equipment which are not intended to be constructed as part of the Project during the Construction Phase.

9.62 **Renderings** services relating to providing graphic pictorial representations, as required by the Owner, of the proposed Project and consisting of:

.01 Black and white elevation view(s)
.02 Black and white perspective view(s)
.03 Elevation view(s) in color
.04 Perspective view(s) in color.

9.63 **Model Construction** services consisting of preparation of:

.01 Small-scale block model(s) showing relationship of structure(s) to site
.02 Moderate-scale block model(s) of structure(s) designed for the Project
.03 Moderate-scale detailed model(s) of structure(s) designed for the Project showing both interior and exterior design
.04 Large-scale models of designated interior or exterior components of the Project.

9.64 **Life Cycle Cost Analysis** services consisting of assessment, on the basis of established relevant economic consequences over a given period of time, of:

.01 A given planning and design solution for the Project
.02 Alternative planning and design solutions for the Project
.03 Selected systems, subsystems or building components proposed for the Project.

9.65 **Value Analysis** services consisting of the review during design phases of the cost, quality and time influences of proposed building materials, systems and construction methods relative to design objectives in order to identify options for obtaining optimum value for the Owner.

9.66 **Quantity Surveys** services consisting of a detailed determination of the quantities of materials to be used in the Project to establish the basis for price determination by bidding or negotiations.

9.67 Detailed Construction Cost Estimates services consisting of:

- **.01** Development, when the construction documents are approximately 90% complete, of a Detailed Estimate of Construction Cost based on quantity take-offs and unit cost pricing for materials, labor, tools, equipment and services required for the Work plus estimates for the Contractor's supervision cost, Work required by General and Supplementary Conditions, and an allowance for a reasonable Contractor's overhead and profit; or
- **.02** Continuous development during all phases of design and documentation, of an Estimate of Construction Cost for the purpose of greater cost control; culminating in a Detailed Estimate of Construction Cost or detailed quantity surveys or inventories of material, equipment and labor.

9.68 Energy Studies services consisting of special analyses of mechanical systems, fuel costs, on-site energy generation and energy conservation options for the Owner's consideration.

9.69 Environmental Monitoring services consisting of:

- **.01** Monitoring of air, water and other designated components of the environment to establish existing conditions, and the preparation of related analyses and reports.

9.70 Tenant-related Services consisting of design and documentation services for tenants or potential tenants for:

- **.01** Space planning, partition and furnishings locations, and furniture and equipment layouts
- **.02** Material and color selections and coordination
- **.03** Adaptation of mechanical, electrical and other building systems to meet tenant needs
- **.04** Statements of Probable Construction Cost.

9.71 Graphics Design services consisting of:

- **.01** Design of interior and exterior signs and identifying symbols
- **.02** Design of special logos or graphics for the Project
- **.03** Documentation of requirements for procurement of graphics work
- **.04** Managing procurement of graphics work
- **.05** Coordination of delivery and installation.

9.72 Fine Arts and Crafts Services relating to acquisition of fine arts or crafts to be a part of the Project and consisting of:

- **.01** Consultations on selection, commissioning and/or execution
- **.02** Design integration
- **.03** Managing procurement
- **.04** Purchasing fine arts or crafts on the Owner's behalf.

9.73 Special Furnishings Design services relating to Architect-designed special furnishings and/or equipment incorporated into or provided for the Project and consisting of:

- **.01** Design and documentation
- **.02** Specifications or standards
- **.03** Management of procurement
- **.04** Coordination of installation
- **.05** Purchase on the Owner's behalf.

9.74 Non-Building Equipment Selection services relating to equipment and furnishings not incorporated into the construction of the Project and consisting of:

- **.01** Establishment of needs and criteria
- **.02** Preparation of requirements, Specifications and bidding or purchasing procedures
- **.03** Management of procurement
- **.04** Coordination of delivery and installation
- **.05** Purchase of non-building equipment on the Owner's behalf.

9.75 Project Promotion/Public Relations services relating to presentation of the Project to the public or identified groups and consisting of:

- **.01** Preparation of press releases
- **.02** Preparation of special brochures and/or promotional pieces
- **.03** Assistance in production and distribution of promotional materials
- **.04** Presentations at public relations and/or promotional meetings.

9.76 Leasing Brochures services relating to preparation of special material to assist the Owner in leasing the Project and consisting of:

- **.01** Design
- **.02** Preparation of illustrations and text
- **.03** Arranging for and managing production.

9.77 Expert Witness services consisting of preparing to serve and/or serving as an expert witness in connection with any public hearing, arbitration proceeding or legal proceeding.

9.78 Computer Applications services consisting of computer program development and/or computer program search and acquisition, plus on-line computer time charges, for:

- **.01** Programming
- **.02** Economic feasibility analysis
- **.03** Financial analysis
- **.04** Site analysis
- **.05** Construction cost estimating
- **.06** Detailed Project scheduling
- **.07** Market analysis
- **.08** Architectural analysis and design
- **.09** Structural analysis and design
- **.10** Mechanical analysis and design
- **.11** Electrical analysis and design
- **.12** Production of Drawings
- **.13** Construction cost accounting.

9.79 Materials and Systems Testing services relating to testing of components of the completed Project for conformance with Contract requirements and consisting of:

- **.01** Establishment of requirements
- **.02** Procurement of testing services
- **.03** Monitoring testing
- **.04** Review, analysis and reporting of testing results.

9.80 Demolition Services consisting of:

- **.01** Preparation of Contract Documents for demolition of existing structures
- **.02** Managing the bidding/negotiation/award process
- **.03** Providing field observation and general administration services during demolition.

9.81 Mock-up Services relating to the construction of full-size or half-size details of components of the Project for study and testing during the design phases and consisting of:

- **.01** Design and documentation for the required mock-up(s)
- **.02** Management and coordination of pricing and contracting for mock-up services
- **.03** Construction administration of mock-up construction activities
- **.04** Arrangements for testing and monitoring performance of mock-up(s)
- **.05** Administration of testing and monitoring services
- **.06** Review, analysis and reporting of results of testing and monitoring services.

9.82 Still Photography services consisting of:

- **.01** Documentation of existing conditions
- **.02** Aerial site photography
- **.03** Photographic recording for study purposes of facilities similar to the Project
- **.04** Periscopic photography of models for the Project
- **.05** Presentation photography of rendering(s) and model(s) for the Project
- **.06** Construction progress photography
- **.07** Architectural photography of the completed Project.

9.83 Motion Pictures and Videotape services relating to preparation of promotional or explanatory presentations of the Project during the design and/or construction phases.

9.84 Coordination with Non-Design Professionals services consisting of consultation and coordination with non-design professionals such as economists, sociologists, attorneys and accountants, for:

- **.01** Preparation of economic studies
- **.02** Condominium documentation
- **.03** Sociological impact studies.

9.85 Special Disciplines Consultation services consisting of retaining, directing and coordinating the work of special disciplines consultants identified from the following list or other sources attached hereto, whose specialized training, experience and knowledge relative to specific elements and features of the Project are required for the Project:

- **.01** Acoustics
- **.02** Audio-Visual
- **.03** CPM Scheduling
- **.04** Code Interpretation
- **.05** Communications
- **.06** Computer Technology
- **.07** Concrete
- **.08** Cost Estimating
- **.09** Demography
- **.10** Display
- **.11** Ecology
- **.12** Economics
- **.13** Editorial
- **.14** Elevators/Escalators
- **.15** Fallout Shelters
- **.16** Financial
- **.17** Fire Protection
- **.18** Food Service
- **.19** Insurance
- **.20** Legal
- **.21** Life Safety
- **.22** Lighting
- **.23** Management
- **.24** Materials Handling
- **.25** Psychology
- **.26** Public Relations
- **.27** Radiation Shielding
- **.28** Real Estate
- **.29** Reprographics
- **.30** Safety
- **.31** Sociology
- **.32** Soils/Foundations
- **.33** Space Programming
- **.34** Specifications
- **.35** Traffic/Parking
- **.36** Transportation.

9.86 Special Building Type Consultation services consisting of retaining, directing and coordinating the work of special building type consultants whose specialized training, experience and knowledge relative to the requirements, planning and design of the Project are required for the Project.

PROJECT:

PROJECT #: **DATE:**

The Owner and the Architect, respectively, shall provide for this Phase of the Project those services appropriately designated and assigned below and described subsequently.

OWNER *(signed)*__________________________

ARCHITECT *(signed)*__________________________

	By Architect	By Architect, as Outside Services			By Owner, coordinated by Architect	By Owner			By Architect, as Additional Service*	Not to be provided	Method of Compensation

METHODS OF COMPENSATION

A = Multiple of Direct Salary Expense
B = Multiple of Direct Personnel Expense
C = Professional Fee Plus Expenses
D = Percentage of Construction Cost
E = Stipulated Sum
F = Hourly Billing Rates
G = Multiple of Amounts Billed to Architect
H = Other:

*Requires separate authorization and compensation

AIA DOCUMENT B162 • SCOPE OF DESIGNATED SERVICES • NOVEMBER 1977 EDITION • AIA® • ©1977
THE AMERICAN INSTITUTE OF ARCHITECTS, 1735 NEW YORK AVENUE, N.W., WASHINGTON, D.C. 20006

ENUMERATION OF MODIFICATIONS TO DESIGNATED SERVICES

PROJECT:

PROJECT #: DATE:

Prior to the execution of this Document B162, modifications of the Descriptions of Designated Services were made only with respect to the following numbered items:

Services which have been deleted in their entirety are indicated on the Schedule of Designated Services pages by marking the column, "Not To Be Provided," and also on the accompanying Description of Designated Services pages by ruling them out.

OWNER

ARCHITECT

INSTRUCTION SHEET *AIA DOCUMENT B162a*

FOR AIA DOCUMENT B162, SCOPE OF DESIGNATED SERVICES—NOVEMBER 1977 EDITION

A. GENERAL INFORMATION:

1. New Documents B161 and B162:

This is the first edition of *AIA Document B162, Scope of Designated Services,* and is intended to be used in conjunction with *AIA Document B161, Owner-Architect Agreement for Designated Services.* These new forms are designed to work together in determining (1) the terms and conditions and the amounts of compensation of the Agreement (B161) and (2) the responsibilities and services to be undertaken by Owner and Architect (B162). The separation of B161 and B162 provides for the flexibility of using B162, the *Scope of Designated Services,* with other agreement forms. Alternatively, B161 may be used as the terms and conditions with other forms of scope of services.

In order to avoid problems of consistency and coordination, especially with regard to the methods of compensation provided for in B161 and applied to specific services in B162, B161 and B162 should be used together whenever possible. Neither B161 nor B162 may be used alone. A complete agreement will require (1) both B161 and B162, (2) B161 with another equivalent to B162, or (3) B162 with another equivalent to B161.

2. AIA Financial Management System:

Documents B161 and B162 are designed to interface with the AIA cost-based approach to compensation as set forth in *Compensation Guidelines for Architectural and Engineering Services* AIA Catalog #M-188, 1978 Edition), and with all elements of the AIA Financial Management System. These publications and practice aids, particularly *Compensation Guidelines,* should be consulted for a full understanding of the process of establishing the scope of services and compensation.

The B161-B162 combination represents a departure from previous AIA Owner-Architect Agreements in that (1) the traditional five phases of services have been expanded to nine categories, enabling predesign, site analysis, postconstruction and supplemental services to be provided under the same terms and conditions as are other services, (2) the traditional distinction between "Basic" and "Additional" services and consequent variations in compensation methods is not recognized in these new documents, and (3) most of the services that could be provided by the Architect are included in B162, *but appropriate editing and modification by the parties prior to execution of the Agreement is required.*

3. Description:

B161 consists of the following elements:

(a) Cover page
(b) Terms and conditions (Article 1 through 13)
(c) Compensation (Article 14)
(d) Other Conditions or Services (Article 15)

B162 consists of the following elements:

(a) Cover page
(b) Nine Phases of Services and one Phase in blank, including:
  (i) Schedules of Designated Services (grid sheets)
  (ii) Descriptions of Designated Services
(c) Enumeration of Modifications page

See the Instruction Sheet for AIA Document B161 (AIA Document B161a) for further information on AIA Owner-Architect Agreements.

B. COMPLETING THE B162 SCOPE OF SERVICES FORM:

1. Cover Page:

Identification of the date, name or title of the project and parties to the Agreement should be made in the same manner as indicated on the Owner-Architect Agreement form (B161).

2. Schedule of Designated Services forms (grid sheets):

These forms, preceding the descriptions of services for each Phase, should be completed only after an understanding has been reached with the Owner on the following items:

(a) Phases of Service and specific services (numbered line-items on each sheet) to be provided (whether by Architect or others)
(b) Assignment of responsibility for services
(c) Description of specific services
(d) Method of compensation applicable to each service or Phase and total amount of compensation

The worksheet forms and instructions incorporated into *Compensation Guidelines for Architectural and Engineering Services* are intended to be completed prior to the use of the B162 forms.

Phases not to be provided: If an entire Phase is not to be provided, the pages containing the schedule and descriptions of services should be retained, in order to provide both parties with a clear record of Phases and services not to be provided.

The Phases in which services are to be provided should be noted in Article 15 of B161, and the Phases which are not to be provided should be stricken out and noted in the Enumeration of Modifications page. Alternatively, if entire Phases of service are not to be provided these can be discarded, but their exclusion from the Agreement should be noted in Article 15 of B161 along with those Phases which are included.

Services not to be provided and identification of parties furnishing services: If particular services within a Phase are not to be provided, the vertical column labeled "Not To Be Provided" should be checked on the line containing that service. Because the Designated Services Agreement contemplates that services will be required of both the Owner and the Architect in order to complete the Project, service identifications should not be stricken simply because they are not to be provided by the Architect. The "Not To Be Provided" column is the appropriate means to designate services which will not be required. Similarly, if the service is to be provided by the Architect (using in-house staff), by the Architect through outside services (consultants), by the Owner without assistance from the Architect, or by the Owner or Owner's consultants but coordinated by the Architect, the appropriate vertical column on each line should be checked. Services to be provided by entities other than parties to this Agreement can be specified by inserting in one or more of the vacant vertical columns an appropriate identification of the entity by which the service will be provided and identification of the party to which that entity is responsible. Identification of different arrangements should be set forth and described in more detail in Article 15 of B161.

When a particular service is to be performed partially by one party and partially by another, or when the responsibility for a particular service is to be shared between two parties, *both* columns should be checked. The *descriptions* of such services following the "grid sheets" should be modified by adding in parentheses the identification of the party (such as "Owner" or "Architect") responsible for the particular service described when a division of responsibility can be made, or when the responsibility for the subservice is *shared,* by listing both parties in parentheses and indicating the extent of the responsibility for each. When it is desirable to identify the primary responsibility in a shared service, the parenthetical notation should so indicate.

Undefined services: Often there will be services for which the need is not clear, which may not be established until a later date, or which are undefined in their extent or in the amount of compensation required. Sometimes the need for such services will not be determined until after completion of one or more Phases of the project or until further consideration by the Owner. Such services may be treated as Additional Services under Paragraph 1.4 of B161, requiring separate authorization and compensation. To designate these services, the "By Architect, as Additional Service" column is provided.

Including Services not described in B162: Service identification titles may be modified by striking out and typing in, as appropriate. Inclusion of services not listed on the schedule is accomplished by typing in the identification of the service below the printed schedule, and checking the appropriate vertical column to assign responsibility for that service.

Adding a new Phase: A new Phase of services can be added to the Agreement by completing the blank Schedule of Services Form (grid sheet) and inserting appropriate definitions of the individual services and assigning page numbers. If not used, this page should be discarded.

Numbering system for services: The numerical designations of the Phases and services have been assigned for consistency with the entire AIA Financial Management System. The numbering system is designed to meet the needs of several components of this system and the logic will not necessarily be apparent to the user of B162. However, it does provide a flexible framework for adding services in preparing B162 and all numbers under each Phase will be in ascending numerical order. Services added to the list should be numbered with unassigned double-digit numbers as suffixes; these currently include 11, 12, 20, 55-60 and 87-99. Use the Phase number as the prefix. Example: If acoustical consultation is to be provided in the Design Development Phase, an appropriate entry would be "4.87 Acoustical Consultation."

B162 introduces an additional refinement by numbering the subservice headings under each service heading for easier identification and referencing. Example: a reference to a modification to "Research of Community Attitudes" under 1.03, Agency Consulting/Review/Approval services, would be 1.03.03.

Compensation Method: The Method under which each service will be compensated, as set forth in Article 14 of B161, is to be indicated by entry of the appropriate letter in the last column of the form. These Methods are described in *Compensation Guidelines.*

Where a stipulated-sum or percentage method of compensation applies, the dollar amounts or percentages set forth in B161 will cover the total compensation for all services in all Phases to be compensated for by that method, unless such compensation is specifically subdivided by Phases or services in B161.

Completion: After the form has been completed, a line should be drawn through all empty spaces of the form. See Exhibit 3-2 in *Compensation Guidelines* for an example. Duplicate copies should then be prepared for execution by the Owner and the Architect in the upper left-hand corner, with identification of the project name, number and date of the Agreement.

3. Descriptions of Designated Services pages:

Descriptions of the services follow immediately after the "grid sheet" for each Phase. The description of any services for which the "Not To Be Provided" column has been checked on the "grid sheet" should be stricken entirely.

Modifications to the service descriptions: These can be accomplished by striking out the portions of the descriptions which are inapplicable, and inserting any additional data within the space provided for that service. Note that some descriptions contain multiple choices, some of which must be deleted. Supplemental services may either be covered on the Phase 9 pages or added to one or more of the appropriate sequential Phases 1 through 8. Similarly, services from Phases 1 and 2 may be transferred, if appropriate, to later Phases. Services not described in B162 and added on the face of the "grid sheet" for any Phase should be described following the descriptions of other services for that Phase. Any services or subservices which have been modified should be indicated on the Enumeration of Modifications page.

4. Enumeration of Modifications:

Enumeration of Modifications to Designated Services constitutes a convenient method for listing the items in the Descriptions of Designated Services which have been modified. This enumeration is verified by the signatures of the Owner and the Architect. An alternate method of verifying modifications, such as the initials or signatures of the Owner and the Architect on each page where a modification has been made, may be used instead.

5. Execution of the Agreement:

Execution of B161 and B162 requires that both parties sign the last page of B161 and B162. Any modifications to the Agreement should be initialed by both parties (see "Enumeration of Modifications page" above). Once duplicate copies of the Agreement have been executed, B161 and B162 should be stapled or otherwise permanently attached to one another.

C. SUBSEQUENT MODIFICATIONS:

Subsequent modifications to the Agreement, such as an addition, deletion, or change in description or assignment of any services or an authorization to perform Additional Services, should be made by a separate instrument referring specifically to the document originally executed. Changes should not be made to B161 or B162 after execution.

D. REPRODUCTION:

AIA Document B162 is a copyrighted document, and may not be reproduced or excerpted from in substantial part without the express written permission of AIA. Purchasers of B162 are hereby entitled to reproduce a maximum of ten copies of the completed or executed document for use only in connection with the particular project. AIA will not permit the reproduction of this document in blank, or the use of substantial portions of, or language from this Document, except upon written request and after receipt of written permission from AIA.

1/78. 2M188 Rev. 1/78